Nanofluid Flow, Heat, and Mass Transfer at Stretching Surfaces

Nanofluid Flow, Heat, and Mass Transfer at Stretching Surfaces explores nanofluid and hybrid nanofluid flow at a moving surface, stretching plate, stretching disk, and past a stretching surface. It demonstrates the wide range of applications of nanofluid heat and mass transport enhancement in many industrial and manufacturing processes.

With a combination of numerical and analytical approaches, the book presents multiple solution strategies, including boundary layer flow problems with dimensionless equations, numerical simulation, and theoretical approaches. It discusses unique physical phenomena, including phase change, buoyancy, chemical reactions, moving/stretching surfaces, and magnetohydrodynamics. The book analyzes hybrid nanofluid flow past a stretching surface in the presence of a magnetic field and the effects of activation energy and heat generation/absorption. It includes real-world case studies and solutions.

The book will interest industry researchers and graduate students studying applications of nanotechnology and nanofluids within the fields of mechanical, chemical, manufacturing, and biomedical engineering.

Nanofluid Flow, Heat, and Mass Transfer at Stretching Surfaces

Kuppalapalle Vajravelu, John P. Abraham,
Swati Mukhopadhyay, and P. Lakshminarayana

CRC Press
Taylor & Francis Group
Boca Raton London New York

CRC Press is an imprint of the
Taylor & Francis Group, an **informa** business

First edition published 2026
by CRC Press
2385 NW Executive Center Drive, Suite 320, Boca Raton, FL 33431

and by CRC Press
4 Park Square, Milton Park, Abingdon, Oxon, OX14 4RN

CRC Press is an imprint of Taylor & Francis Group, LLC

ISBN: 978-1-032-94532-3 (hbk)
ISBN: 978-1-032-94533-0 (pbk)
ISBN: 978-1-003-57127-8 (ebk)

DOI: 10.1201/9781003571278

Typeset in Times
by Newgen Publishing UK

Contents

Preface

This book is intended to be read by scientists and engineers, both in industry and in academia. For students or working professionals in the thermal sciences, this text provides a solid foundation for nanoparticle-embedded fluid flows. We comprehensively explore a variety of nanofluids in various heat and mass transport situations. These nanofluids, which are created by suspending nanoparticles within a base fluid, show tremendous promise for improving heat transfer. They have emerged as revolutionary materials due to their improved thermal conductivity and heat transport performance. We identify numerous application areas, including biomedical systems, manufacturing processes, electronics cooling, and energy systems.

The first chapter lays the foundation for foundational concepts regarding nanofluids, nanoparticle stabilization, and optimization. Subsequent chapters analyze a series of increasingly complex real-world applications, including flow across various geometrical objects such as stretching surfaces, moving surfaces, rotating bodies, wedges, and peristaltic channels. Each chapter examines the interplay of complex physical processes, including magnetic fields (MHD effects), chemical reactions, Brownian motion, thermophoresis, viscous dissipation, and radiation.

The real-world problems are solved using well-established numerical techniques that include similarity transformations, boundary layer approximations, and commercially available software platforms that make these problems accessible to all. The numerical solutions are validated against previously published research to establish the accuracy of the methods.

With this text, scientists and engineers will develop a deep understanding of the physical processes that are present and the solution strategies. Readers will also appreciate the implications of nanofluid applications in the thermal sciences. We hope that readers will be armed with the knowledge and motivation to incorporate nanofluid technology in their work as they optimize case-specific heat transfer and fluid flow processes.

The work in this book was entirely written by the authors. Generative AI was used to summarize salient points in Chapter 9, which was then entirely rewritten by the authors. The generative AI program was ChatGPT 4.5, and it was used to expedite the organization of Chapter 9. No other chapters involved generative AI.

Acknowledgements

Swati Mukhopadhyay acknowledges the help received from her students Hiranmoy Maiti, Sudip Dey, and Saheb Konai for drawing figures of Chapters 3, 5, 6, and 8.

P. Lakshminarayana acknowledges the help received from his students K. Thirunavukarasan, Ilango M.S, P. Yamuna Rani and Ramki R for drawing the figures of Chapters 1, 2, 4, and 7.

About the Authors

Kuppalapalle Vajravelu has obtained a Ph.D. degree in Applied Mathematics from the Indian Institute of Technology, Kharagpur, India. He is a professor at the University of Central Florida, Orlando, Florida, USA. He has been actively engaged in research in the fields of Nanofluid Flow, Biofluids, Fluid Dynamics, and Numerical Heat Transfer for the past 47 years. Several students did Undergraduate Research with him and got NSF fellowships to do Ph.D. Furthermore, he supervised M.S. and Ph.D. students, and published several (450 plus) research articles and books (five) on Differential Equations and their Applications. Also, he reviewed papers for reputed national and international journals. He has a total of 15,936 citations, with an H-index of 64 (as per Google Scholar). He was the Editor-in-Chief of the journal *Differential Equations and Nonlinear Mechanics* for three years. Also, he is on the editorial boards of several journals.

John P. Abraham is a professor of mechanical engineering at the University of St. Thomas in Minnesota, USA. His research and teaching activities relate to the mechanical engineering discipline and include design with a special focus on the thermal sciences. He has published approximately 500 works, including journal papers, book chapters, books, conference presentations, and patents. He is an active consultant and has worked in the medical device, defense, sustainability, manufacturing, and other industries.

Swati Mukhopadhyay is working at the University of Burdwan, Burdwan, West Bengal, India. She obtained a Ph.D. degree from the same university and has been actively engaged in research in the field of Fluid Dynamics, Nanofluid Flow, and Heat Transfer for the past 18 years. She published more than 180 research articles and a book. A number of students obtained their Ph.D. degrees under her able supervision. She has 7,454 citations with h index: 49 (by Google Scholar) and 5,553 citations with h index: 43 (by SCOPUS) as of 26 April, 2025.

P. Lakshminarayana is working at Vellore Institute of Technology, Vellore, Tamil Nadu, India. He obtained a Ph.D. degree from Sri Venkateswara University, Tirupati, Andhra Pradesh, India. He is an active researcher in the field of Fluid Dynamics, Peristaltic Transport, Boundary Layer, Nanofluid Flow, and Heat and Mass Transfer for the past 15 years. He has published more than 75 research articles and three book chapters. Three students obtained their Ph.D. degrees, and five students are pursuing their Ph.D., under his supervision. He has 2,126 citations with h index: 28 (by Google Scholar) and 1,746 citations with h index: 26 (by SCOPUS) as of 27 April, 2025.

1 Introduction to Nanofluids and Hybrid Nanofluids

1.1 BACKGROUND

Nanotechnology, particularly nanofluids, is an innovative collection of technologies that has the potential to significantly impact various scientific and technical domains. Fluids enhanced with nanoparticles (nanofluids) have exceptional qualities capable of increasing heat transfer, optimizing effectiveness, and maximizing performance in a variety of applications, such as electronics, energy generation, and cooling systems. This chapter provides a comprehensive overview of the most recent findings from nanofluid research, emphasizing significant developments, challenges, and practical applications. It also aims to identify knowledge gaps and provide possible directions for further research and advancement in this rapidly developing sector. This chapter facilitates the further integration of nanofluid technology into scientific and industrial processes by outlining a route for future innovation and advancement.

Nanofluids are specialized liquids created by dispersing nanoparticles ranging between 1 and 100 nanometers within a base fluid. The notion of nanofluids emerged in the early 1990s, and since then, it has received significant attention. The primary reason for their adoption is their ability to substantially enhance heat transmission with enhanced thermal conductivity – significantly improving its heat conduction relative to conventional fluids.

In 1995, at the United States' Argonne National Laboratory, Choi and Eastman (1995) first proposed the concept of nanofluids. To improve the heat transfer properties of base fluids like water and oil, they suggested incorporating nanoparticles made of metals, ceramics, or carbon-based compounds. Lee et al. (1999) examined the measurement of thermal conductivity in nanofluids that contain oxide nanoparticles (Al_2O_3 and CuO). Their experimental results indicate that these nanofluids exhibit significantly enhanced thermal conductivities compared to the same fluids without nanoparticles. Xuan and Li (2000) conducted a study focusing on the enhancement of thermal transfer in nanofluids. The investigation revealed that a 4% concentration of nanoparticles in the system enhances the thermal conductivity of the nanofluid (*Cu*/water) by 24%.

In 2006, Buongiorno (2006) created a mathematical model to examine convective thermal transfer in nanofluids. His model emphasized two essential mechanisms, Brownian diffusion and thermophoresis, that improve heat transfer in nanofluids.

DOI: 10.1201/9781003571278-1

Brownian diffusion describes the unpredictable motion of nanoparticles caused by thermal energy, which enhances heat distribution. In contrast, thermophoresis involves the movement of nanoparticles driven by temperature gradients, which further facilitates heat transfer. He showed that these mechanisms produce a slip effect between the nanoparticles and the liquids, improving convective heat transfer. Since then, numerous studies have found that nanofluids generally have greater thermal conductivity compared to base fluids (Abdullah et al., 2019; Tielke et al., 2021; and Abutaleb and Imran, 2021).

Hybrid nanofluids represent a novel category of advanced heat transfer fluids that integrate the advantages of various nanoparticles to attain superior thermal performance. The dispersion of two distinct types of nanoparticles, for example, metallic and non-metallic materials, into a base fluid improves thermal conductivity and stability through synergistic effects. The dual-particle approach enhances heat transfer efficiency while efficiently reducing issues such as sedimentation and viscosity commonly encountered in single-nanoparticle fluids. Hybrid nanofluids have the potential to significantly impact various industries, including renewable energy systems, electronics cooling, and biomedical applications, by providing enhanced and efficient solutions for heat management across multiple technological domains. As one example, Dolui et al. (2023) examine the effects of hybrid nanoparticles (Cu–TiO_2) on two-dimensional peristaltic blood flow within a non-uniform cylindrical annulus, focusing on its application in drug delivery. Vajravelu et al. (2024) explored developments in nanofluid flow, heat, and mass transfer at surfaces that are moving or stretching. Figure 1.1 illustrates the nanofluid and hybrid nanofluid models (Alsayegh and Ali, 2020).

1.2 PREPARATION OF NANOFLUIDS

As mentioned earlier, nanofluids are produced by dispersing nanoparticles in a base fluid, such as water or oil. The nanoparticles are carefully distributed uniformly within the fluid to improve its thermal characteristics. Various techniques are used to maintain a uniform distribution of particles and to inhibit their dispersion or aggregation. Additionally, stabilizers are used to enhance the fluid's long-term stability.

1.2.1 Choosing the Base Fluid

Selecting a base fluid that is suitable for the application is the first stage in the process. Water, oils, or glycol-based fluids are standard options. These fluids are selected for their thermal qualities, their hydrodynamic properties, and their compatibility with the system.

1.2.2 Selecting Nanoparticles

Nanoparticles are selected based on their composition, which includes metals such as copper or aluminum, metal oxides like copper oxide or silica, and carbon materials such as graphene. The material properties of these particles are essential as they have a direct impact on the thermal performance of the nanofluid.

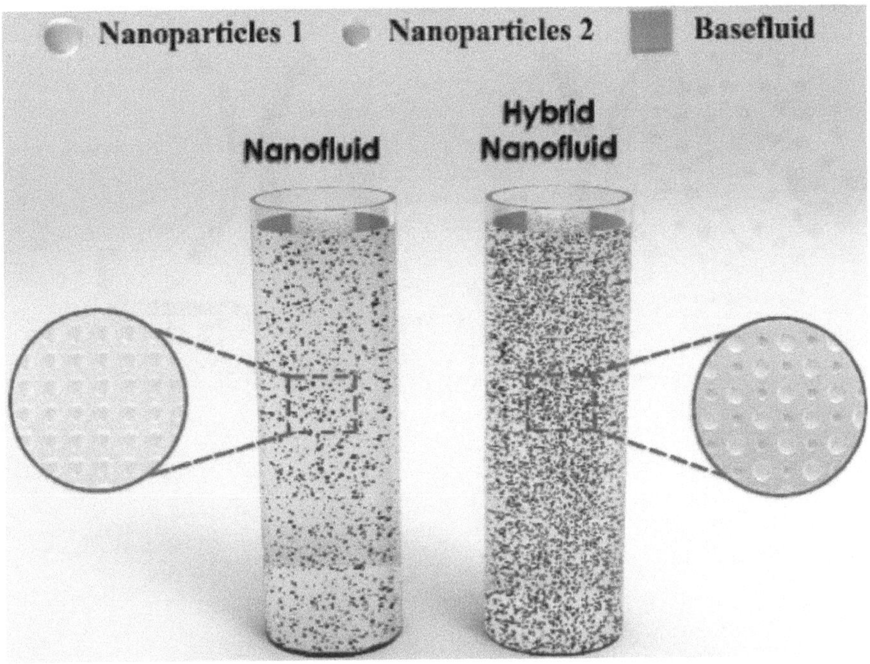

FIGURE 1.1 Nanofluids and hybrid nanofluids. (Adapted from Alsayegh and Ali, 2020).

1.2.3 Dispersion of Nanoparticles

Two primary techniques exist for the dispersion of nanoparticles within the base fluid, which will now be discussed.

1.2.3.1 One-step Method: In the one-step method, nanoparticles are synthesized directly within the base fluid through the application of chemical reactions. This approach facilitates enhanced dispersion and stability of the nanoparticles. The generation of particles within the fluid yields a more uniform distribution throughout the medium. Furthermore, this methodology reduces the likelihood of particle clumping or agglomeration (Zhu et al., 2004; Kumar et al., 2009; and Lee et al., 2012). The one-step process for producing nanofluid is shown in Figure 1.2 (Rabbi et al., 2021).

1.2.3.2 Two-step Method: The two-step method is widely recognized as the primary approach for producing nanofluids. Initially, nanoparticles are produced, typically utilizing either chemical or physical techniques. Subsequently, these particles are incorporated into the base fluid. To achieve an even distribution of nanoparticles within the fluid and prevent clumping, methods such as mechanical stirring, ultrasonic waves (sonication), or magnetic stirring are employed. These techniques contribute to obtaining a uniform, steady, and evenly distributed nanofluid (Suresh et al., 2011; Septiadi et al., 2018; and Sanjeevi and Loganathan, 2020). Figure 1.3 illustrates the two-step method for creating a nanofluid (from Arora and Gupta, 2021).

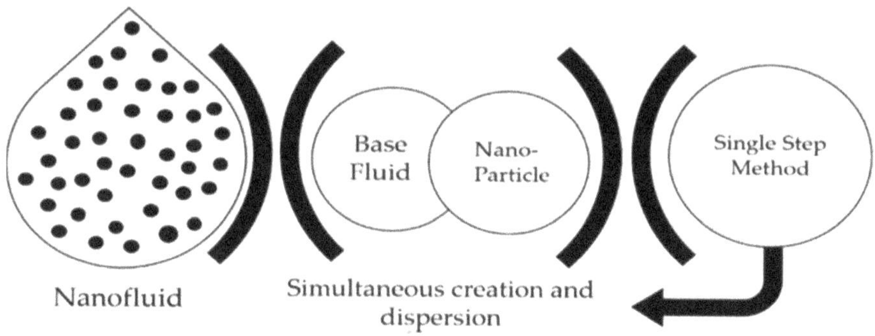

FIGURE 1.2 The one-step method. (Adapted from Rabbi et al., 2021).

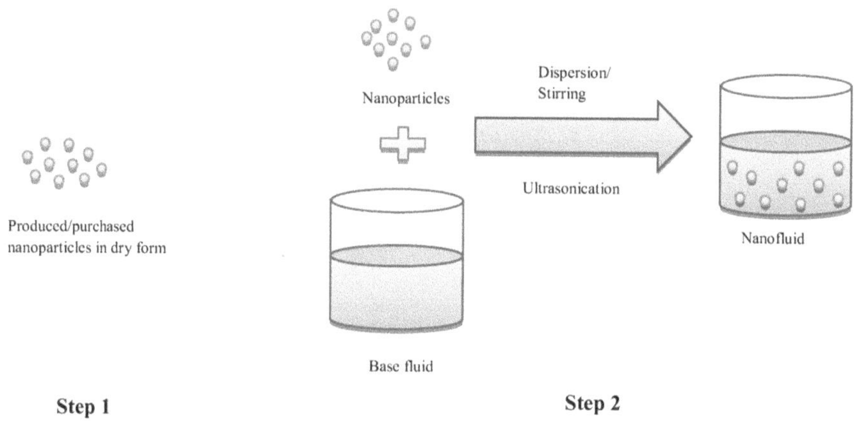

FIGURE 1.3 Illustration of the two-step method. (Adapted from Arora and Gupta, 2021).

1.2.4 Stabilization

The process of stabilization involves maintaining a uniform dispersion of nanoparticles and the prevention of agglomeration or settling. Stabilization is crucial for preserving the intended thermal and fluid characteristics of the nanofluid. The primary methods for stabilizing nanofluids are outlined as follows:

1.2.4.1 Surfactants: Specific chemicals are incorporated to reduce the surface tension and, thus, the aggregation of nanoparticles and facilitate their distribution within the fluid.

1.2.4.2 Ultrasonication: Very high-pitched sound waves are used to break up any nanoparticle groups and mix them into the stream.

1.2.4.3 Electrostatic Stabilization: Charging the nanoparticles prevents them from clumping together by making them repel one another.

1.2.4.4 Steric Stabilization: A protective layer, such as a polymer, is applied to the nanoparticles to prevent them from aggregating.

1.2.4.5 Control pH and Temperature: The stability and uniform mixing of the nanoparticles can be maintained by varying the temperature and pH of the fluid.

1.2.5 Characterization

Characterization refers to a description of the physical and geometric features of a nanofluid. The characterization involved in nanofluid preparation is crucial as it ensures optimal performance and long-term stability of the fluid. Here, we provide an overview of characterization and its significance:

1.2.5.1 Particle Size and Distribution: Assessing the dimensions and distribution of nanoparticles is crucial for achieving uniform dispersion within the fluid. These both have a direct effect on the performance of the nanofluid.

1.2.5.2 Zeta Potential: Zeta potential refers to the surface charge of nanoparticles within the fluid. This quantity is related to the nanofluid's stability. A high zeta potential signifies effective dispersion and stability.

1.2.5.3 Viscosity: Understanding the viscosity of the nanofluid is crucial since it influences the fluid's flow characteristics inside pipes or systems, particularly in cooling applications.

1.2.5.4 Thermal Conductivity: Thermal conductivity refers to the ability of the nanofluid to transfer thermal energy in the absence of advection. Since the purpose of using nanoparticles is to enhance heat transmission, it is essential to evaluate this characteristic.

1.2.5.5 Transmission Electron Microscopy (TEM): Under a microscope, TEM assists in examining the size and shape of nanoparticles to make sure they are consistent and evenly dispersed throughout the fluid.

1.3 GENERAL DISCUSSION OF NANOFLUIDS

Nanoparticles ranging from 1 to 100 nm in size exhibit distinctive characteristics like a high surface area, increased reactivity, and quantum effects. These factors render them suitable for a wide array of applications, including nanofluids. Metallic nanoparticles such as copper, silver, and gold are often used in heat transfer applications because of their superior thermal conductivity. On the other hand, metal oxide nanoparticles like titanium dioxide, zinc oxide, and iron oxide offer stability, UV protection, and magnetic characteristics. Carbon-based nanoparticles, including carbon nanotubes (CNTs) and graphene, exhibit remarkable conductivity, strength, and flexibility, which make them valuable for applications related to energy storage, electronics, and enhanced heat transfer. Ceramic nanoparticles (e.g., SiO_2, Al_2O_3) offer hardness and thermal stability, while polymer nanoparticles (e.g., PS, PE) are biodegradable and are often used in drug delivery. The dispersion of these nanoparticles in fluids such as water or oils leads to enhancements in properties like thermal conductivity and electrical conductivity. This results in improved heat dissipation which is beneficial in applications including electronics, automotive cooling, and biomedical devices.

Nanoparticles can have various shapes, including spherical, cylindrical, platelet, blade, and laminar. Each shape influences the properties of the fluid. Spherical nanoparticles ease dispersion and improve stability within fluids; however, their reduced surface area limits their effectiveness in enhancing heat transfer. Cylindrical nanoparticles exhibit an increased surface area, facilitating improved heat transfer; however, their tendency to aggregate can diminish their overall effectiveness. Platelet- and blade-shaped nanoparticles exhibit a significant surface area increase that enhances heat transfer. Yet their tendency to aggregate or form clumps can lead to stability challenges and an increase in the fluid's viscosity. Laminar nanoparticles possess a layered structure that contributes to their remarkable heat transfer properties; however, their tendency to agglomerate can lead to stability issues. The need to optimize heat transfer while preserving the fluid's stability and flow characteristics fundamentally influences the selection of nanoparticle shape. Illustrations of these different shapes will be provided shortly.

Moreira et al. (2013) examined the physical properties and thermal diffusivity of nanofluids that contain extremely low concentrations of silver or gold nanoparticles. Thermal lens spectroscopy was employed with a dual-beam configuration to conduct the thermal measurements. Enhancements of 20% and 16% in the thermal diffusivity were noted for silver and gold nanofluids, respectively, when compared to pure water. The study by Toghraie et al. (2016) investigated the impact of temperature and nanoparticle concentration on the thermal conductivity of hybrid nanofluids. In their experiment, temperatures ranging from 25 to 50°C were recorded for volume fractions ranging from 0 to 3.5%. According to the experiments, nanoparticles improve heat transfer by augmenting the fluid's thermal conductivity owing to their superior heat-conducting characteristics. Their extensive surface area enhances heat absorption and transfer.

Mekheimer et al. (2018) investigate the peristaltic blood flow utilizing gold nanoparticles as a tertiary nanofluid in catheters with applications in cancer treatment. Arif et al. (2023) conducted a study on the implementation of a blood-based ternary hybrid nanofluid that increased heat transfer rates by 8% and spherical-shaped Fe_3O_4 nanoparticles that contributed to a heat transfer enhancement of up to 4.6%. Platelet-shaped *(Zn)* nanoparticles achieve an increase of up to 9%. Additionally, cylindrical-shaped gold nanoparticles provide an enhancement of up to 10.4%.

A critical factor contributing to the success of a nanofluid is the uniform dispersion of nanoparticles within the base fluid, which is vital for preserving their distinct thermal properties over time. Nanoparticles inherently tend to agglomerate as a result of van der Waals forces and electrostatic interactions. However, advanced methods, including surface modifications, surfactant addition, and ultrasonication, have been effectively utilized to attain a uniform dispersion, thereby ensuring long-term stability. Despite the challenges related to cost-effectiveness, scalability, and environmental impact, continuous advancements in these domains are facilitating the development of more sustainable and commercially viable solutions. The ongoing

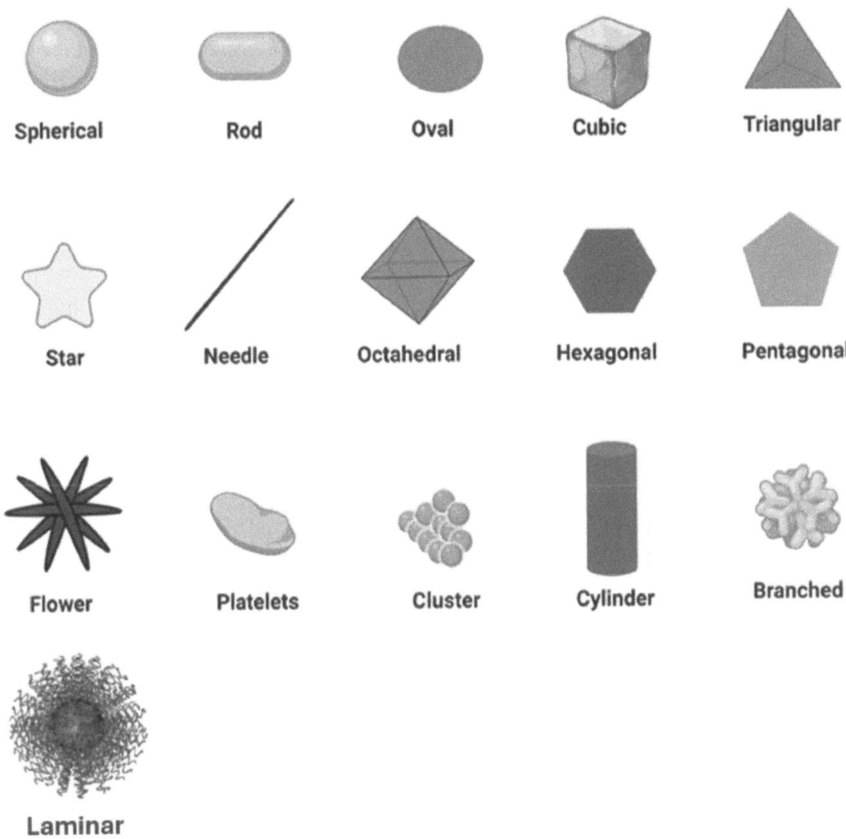

FIGURE 1.4 Different shapes of nanoparticles. (Adapted from Hamida et al., 2020; Thirunavukarasan and Sucharitha, 2025).

research and innovation indicate that the widespread adoption of nanofluids in thermal systems is becoming increasingly viable (Awais et al., 2020; Abo-Elkhair et al., 2021; Elsaid et al., 2021; Khashi'ie et al., 2022; and Afsar et al., 2023). Figure 1.4 illustrates the various shapes of nanoparticles (Hamida et al., 2020; Thirunavukarasan and Sucharitha, 2025).

1.4 NANOFLUID PROPERTIES

Nanofluids demonstrate distinct properties relative to traditional fluids due to the incorporation of nanoparticles inside a base fluid. The following are some of the primary characteristics of nanofluids:

1.4.1 Thermal Conductivity

A key characteristic of nanofluids is their superior thermal conductivity relative to the base fluid. The nanoparticles significantly improve the heat transfer properties of the base fluid, attributed to their elevated surface-area-to-volume ratio and exceptional thermal conductivity. The enhancement of thermal conductivity is influenced by various factors, including the type of nanoparticle material, its concentration, shape, and size. This improvement is the result of the interaction between the base fluid and the nanoparticles, which facilitates the transfer of energy from the base fluid molecules to the nanoparticles, thereby improving heat transport.

1.4.2 Viscosity

The incorporation of nanoparticles into the base fluid often results in enhanced viscosity. The rise in viscosity often results from the interaction between nanoparticles and fluid molecules, resulting in additional flow resistance. The viscosity can be influenced by the concentration, size, shape, and surface chemistry of the nanoparticles. Increases in viscosity can be problematic because it typically means there is an additional pressure drop for flowing fluids. Consequently, more pump energy must be input to generate adequate flows. Thus, it is common for the nanofluids to have increased heat transfer and decreased flow performance, and these two issues must be balanced.

1.4.3 Viscoelasticity

Viscoelasticity in nanofluids pertains to their dual behavior of viscosity and elasticity, indicating that they can flow similarly to liquids while also demonstrating specific solid-like properties at low stress. Nanofluids can display non-Newtonian characteristics, demonstrating either shear-thinning behavior, where the viscosity decreases with an increase in shear rate, or shear-thickening behavior, where viscosity increases with an increase in shear rate. The outcome is dependent upon variables such as the concentration, size, and shape of the nanoparticles. The interactions between nanoparticles and the base fluid influence the flow characteristics of the nanofluid.

1.4.4 Improved Heat Transfer

Nanofluids enhance heat transmission owing to the elevated thermal conductivity of nanoparticles and their augmented surface area. The nanoparticles improve thermal conductivity and promote more efficient energy transfer within the base fluid. Furthermore, their very small size and Brownian motion facilitate efficient heat dispersion, rendering nanofluids more effective than traditional fluids in applications such as cooling systems and heat exchangers.

1.4.5 Increased Density

Nanofluids commonly have a greater density than the base fluid owing to the incorporation of nanoparticles. This characteristic has an impact on the flow and pressure in nanofluid-using systems.

1.4.6 Optical Properties

The optical properties of nanofluids are determined by their size, shape, concentration, and material composition. Their enhanced light absorption properties make them ideal for solar thermal applications, particularly when combined with metallic nanoparticles such as copper or gold. They demonstrate light scattering, which is advantageous for optical devices. By altering the content, size, and presence of metal-based nanoparticles in nanofluids, one can enhance their ability to absorb and scatter light, making them valuable for optical applications and solar energy collection.

1.4.7 Electrical Conductivity

The electrical conductivity of nanofluids is influenced by the type and concentration of nanoparticles present. Specific nanoparticles, particularly metallic types such as copper, can enhance the electrical conductivity of fluids, whereas non-metallic nanoparticles like silica may exert minimal influence.

In general, nanofluids have properties that make them suitable options for many heat transfer and thermal management tasks. Compared to regular fluids, they are possibly more efficient, perform better, and last longer. However, further research is necessary to fully comprehend and enhance the properties of nanofluids for specific applications. We also need to address issues such as stability, scalability, and affordability.

1.5 APPLICATIONS

Nanofluids, consisting of colloidal suspensions of nanoparticles in a base fluid, have garnered significant attention owing to their extensive array of potential applications across various industries. Their outstanding thermal properties and enhanced heat transfer performance position them as formidable solutions for addressing heat transfer challenges and increasing the efficiency of thermal systems. The subsequent sections will delve into a more detailed examination of these applications. Figure 1.5 represents the numerous applications of nanofluids (Shen et al., 2022).

1.5.1 Electronics Cooling

Heat management has grown increasingly complex as electrical components get smaller and become more powerful. Nanofluids are utilized in these devices to enhance cooling. They keep the heat-generating components cool by facilitating more effective heat transfer. Nanofluids are particularly helpful in cooling systems such as heat sinks, which are metallic structures that absorb and release heat, and microchannels, which are tiny tubes that deliver coolant. These cooling systems are frequently found in high-speed processors, LED lights, and computers, where controlling heat is essential for lifespan and optimum performance. As an example, Rafati et al. (2012) studied computer cooling methods using nanofluids.

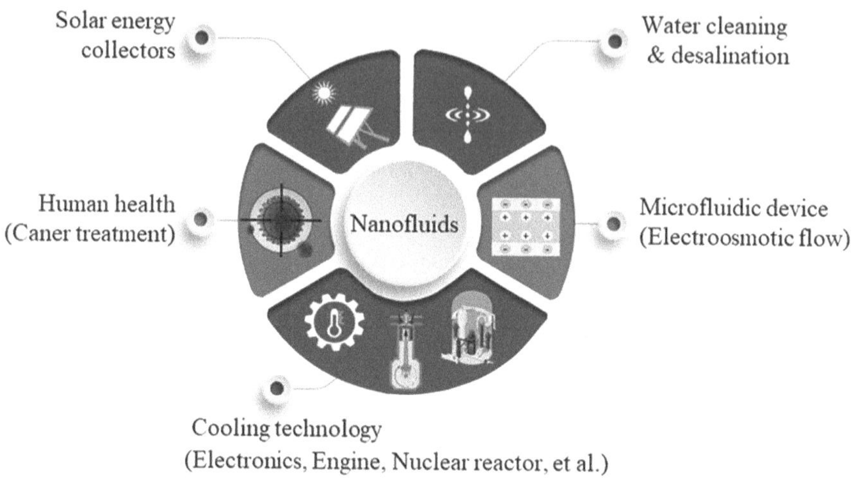

FIGURE 1.5 Applications of nanofluids. (Adapted from Shen et al., 2022).

1.5.2 Heat-transfer Systems

Nanofluids are applicable in heat exchangers, significantly enhancing the efficiency of heat transfer. The incorporation of nanoparticles into a base fluid enhances the thermal conductivity of the fluid, thereby facilitating more effective heat transfer. Nanofluids are particularly advantageous in sectors such as power generation, refrigeration, and chemical processing, where efficient heat management is critical for maintaining optimal system performance. Sajid and Ali (2019) conducted a critical evaluation of current advancements in the application of nanofluids in heat transfer devices.

1.5.3 Solar Energy Systems

Nanofluids are being utilized more and more in solar thermal collectors to improve the efficiency of solar water heating systems. The nanoparticles are better at absorbing and holding heat compared to traditional fluids, which enhances the energy absorption and heat transfer from solar collectors to the working fluid. This results in increased temperatures and improved efficiency in generating solar power. As one example, Colangelo et al. (2016) examined the thermal conductivity, viscosity, and stability of Al_2O_3-based diathermic oil nanofluids for their potential application in solar energy systems.

1.5.4 Automotive Cooling Systems

In vehicle cooling systems, nanoparticles can be added to water, antifreeze, or other liquids. The fluid's capacity to transport heat is enhanced by these nanoparticles, increasing the cooling system's effectiveness. Controlling engine temperature enhances performance and prolongs the engines lifespan. Often, the goal of research in this area is to determine the ideal nanoparticle size and quantity while maintaining the fluid's stability and efficacy. Leong et al. (2010) conducted research on the efficiency of a vehicle radiator that used coolants composed of nanofluids.

1.5.5 Biomedical Devices

Nanofluids are being investigated for their possible use in biomedical devices due to their ability to increase heat transfer and provide superior cooling. They can be utilized in artificial organs, cancer treatment, and medical imaging systems to properly regulate temperature, minimize overheating, deliver therapeutic treatments, and improve device efficiency. They can also be used for thermal management of implanted devices. Nanofluids could also be employed in drug delivery systems to convey medicine to specific bodily areas better. Further research is being conducted to further understand their potential benefits. Examples of relevant research in this area include Dib et al. (2014), who investigated the effect of balloon inflation pressure on cell viability using single and multi-lumen catheters, and Plourde et al. (2014) who investigated intracoronary administration of medication with multi-lumen injection catheters. Relevant research publications in the field. More recently, Sheikhpour et al. (2020) examined the significance of nanofluids in the realms of drug delivery and biomedical technology, focusing on various methods and applications. Other representative uses of nanofluids in biomedical applications include: (Stark et al., 2013; Abraham et al., 2013; and Abraham et al., 2013).

1.5.6 HVAC Systems

Nanofluids improve the efficiency of heat exchangers, coolers, and heaters in HVAC (Heating, Ventilation, and Air Conditioning) systems by transporting heat more readily, resulting in reduced energy use. HVAC systems can thus run more efficiently and consume less energy. However, there are certain drawbacks, such as greater prices, the necessity to ensure that the nanoparticles are uniformly distributed in the fluid, and the possibility that the particles will cause damage to specific system components. Overall, employing nanofluids can make HVAC systems more energy-efficient and beneficial to the environment. Firouzfar et al. (2011) conducted a study to investigate the methanol-silver nanofluid-filled thermosyphon heat exchanger, evaluating its effectiveness and energy savings compared to pure methanol. According to the experiments, using a methanol-silver nanofluid can save anywhere from 8.8% to 31.5% of energy when cooling the supply air stream in an air conditioning system.

REFERENCES

Abdullah, A., Mohamad, I.S., Hashim, A.Y.B. and Md, S.H.S., 2019. Study of thermal conductivity and heat transfer coefficient of mwcnt-oh/deionized water-ethylene glycol based nanofluids. *defence S&T Technical Bulletin*, *12*(2), pp.226–237.

Abo-Elkhair, R.E., Bhatti, M.M. and Mekheimer, K.S., 2021. Magnetic force effects on peristaltic transport of hybrid bio-nanofluid (AuCu nanoparticles) with moderate Reynolds number: An expanding horizon. *International Communications in Heat and Mass Transfer*, *123*, p.105228. https://doi.org/10.1016/j.icheatmasstransfer.2021.105228

Abraham, J., Stark, J., Gorman, J., Sparrow, E. and Kohler, R., 2013. A model of drug deposition within artery walls. *Journal of Medical Devices*, *7*(2), p.020902.

Abraham, J.P., Gorman, J.M., Sparrow, E.M., Stark, J.R. and Kohler, R.E., 2013. A mass transfer model of temporal drug deposition in artery walls. *International Journal of Heat and Mass Transfer*, *58*(1–2), pp.632–638. http://dx.doi.org/10.1016/j.ijheatmasstransfer.2012.11.053

Abutaleb, A. and Imran, M., 2021. Thermal conductivity enhancement for CuO nanoflakes in oil-based and oil blend-based nanofluids. *Journal of the Chinese Chemical Society*, *68*(8), pp.1400–1404. https://doi.org/10.1002/jccs.202100005

Afsar, H., Peiwei, G., Alshamrani, A., Alam, M.M., Hendy, A.S. and Zaky, M.A., 2023. Electroosmotically induced peristaltic flow of a hybrid nanofluid in asymmetric channel: Revolutionizing nanofluid engineering. *Case Studies in Thermal Engineering*, *52*, p.103779. http://doi.org/10.1016/j.csite.2023.103779

Alsayegh, A. and Ali, N., 2020. Gas turbine intercoolers: Introducing nanofluids—A mini-review. *Processes*, *8*(12), p.1572. https://doi.org/10.3390/pr8121572

Arif, M., Di Persio, L., Kumam, P., Watthayu, W. and Akgül, A., 2023. Heat transfer analysis of fractional model of couple stress Casson tri-hybrid nanofluid using dissimilar shape nanoparticles in blood with biomedical applications. *Scientific Reports*, *13*(1), p.4596. https://doi.org/10.1038/s41598-022-25127-z

Arora, N. and Gupta, M., 2021. Thermo-hydraulic performance of nanofluids in enhanced tubes-A review. *Heat and Mass Transfer*, *57*, pp.377–404. https://doi.org/10.1007/s00231-020-02958-y

Awais, M., Shah, Z., Perveen, N., Ali, A., Kumam, P., Rehman, H.U. and Thounthong, P., 2020. MHD effects on ciliary-induced peristaltic flow coatings with rheological hybrid nanofluid. *Coatings*, *10*(2), p.186. https://doi.org/10.3390/coatings10020186

Buongiorno, J., 2006. Convective transport in nanofluids. *Journal of Heat Transfer*, 128(3), pp.240–250. https://doi.org/10.1115/1.2150834

Choi, S.U. and Eastman, J.A., 1995. *Enhancing thermal conductivity of fluids with nanoparticles* (No. ANL/MSD/CP-84938; CONF-951135-29). Argonne National Lab. (ANL), Argonne, IL (United States).

Colangelo, G., Favale, E., Miglietta, P., Milanese, M. and de Risi, A., 2016. Thermal conductivity, viscosity and stability of Al2O3-diathermic oil nanofluids for solar energy systems. *Energy*, *95*, pp.124–136. https://doi.org/10.1016/j.energy.2015.11.032

Dib, N., Schwalbach, D.B., Plourde, B.D., Kohler, R.E., Dana, D. and Abraham, J.P., 2014. Impact of balloon inflation pressure on cell viability with single and multi lumen catheters. *Journal of Cardiovascular Translational Research*, *7*, pp.781–787. https://doi.org/10.1007/s12265-014-9595-z

Dolui, S., Bhaumik, B., De, S. and Changdar, S., 2023. Effect of a variable magnetic field on peristaltic slip flow of blood-based hybrid nanofluid through a nonuniform annular channel. *Journal of Mechanics in Medicine and Biology*, *23*(01), p.2250070. https://doi.org/10.1142/S0219519422500701

Elsaid, E.M. and Abdel-Wahed, M.S., 2021. Impact of hybrid nanofluid coolant on the boundary layer behavior over a moving cylinder: Numerical case study. *Case Studies in Thermal Engineering*, *25*, p.100951. http://dx.doi.org/10.1016/j.csite.2021.100951

Firouzfar, E., Soltanieh, M., Noie, S.H. and Saidi, S.H., 2011. Energy saving in HVAC systems using nanofluid. *Applied Thermal Engineering*, *31*(8–9), pp.1543–1545. http://dx.doi.org/10.1016/j.applthermaleng.2011.01.029

Hamida, R.S., Ali, M.A., Redhwan, A. and Bin-Meferij, M.M., 2020. Cyanobacteria–a promising platform in green nanotechnology: A review on nanoparticles fabrication and their prospective applications. *International Journal of Nanomedicine*, *15*, pp.6033–6066.

Khashi'ie, N.S., Arifin, N.M. and Pop, I., 2022. Magnetohydrodynamics (MHD) boundary layer flow of hybrid nanofluid over a moving plate with Joule heating. *Alexandria Engineering Journal, 61*(3), pp.1938–1945. http://dx.doi.org/10.1016/j.aej.2021.07.032

Kumar, S.A., Meenakshi, K.S., Narashimhan, B.R.V., Srikanth, S. and Arthanareeswaran, G.J.M.C., 2009. Synthesis and characterization of copper nanofluid by a novel one-step method. *Materials Chemistry and Physics, 113*(1), pp.57–62. https://doi.org/10.1016/j.matchemphys.2008.07.027

Lee, G.J., Kim, C.K., Lee, M.K., Rhee, C.K., Kim, S. and Kim, C., 2012. Thermal conductivity enhancement of ZnO nanofluid using a one-step physical method. *Thermochimica Acta, 542*, pp.24–27. https://doi.org/10.1016/j.tca.2012.01.010

Lee, S., Choi, S.S., Li, S.A. and Eastman, J.A., 1999. *Measuring thermal conductivity of fluids containing oxide nanoparticles. ASME Journal of Heat Transfer*, 121, pp. 280–289. https://doi.org/10.1115/1.2825978

Leong, K.Y., Saidur, R., Kazi, S.N. and Mamun, A.H., 2010. Performance investigation of an automotive car radiator operated with nanofluid-based coolants (nanofluid as a coolant in a radiator). *Applied Thermal Engineering, 30*(17–18), pp.2685–2692. http://dx.doi.org/10.1016/j.applthermaleng.2010.07.019

Mekheimer, K.S., Hasona, W.M., Abo-Elkhair, R.E. and Zaher, A.Z., 2018. Peristaltic blood flow with gold nanoparticles as a third grade nanofluid in catheter: Application of cancer therapy. *Physics Letters A, 382*(2–3), pp.85–93. https://doi.org/10.1016/j.physleta.2017.10.042

Moreira, L.M., Carvalho, E.A., Bell, M.J.V., Anjos, V., Sant'Ana, A.C., Alves, A.P.P., Fragneaud, B., Sena, L.A., Archanjo, B.S. and Achete, C.A., 2013. Thermo-optical properties of silver and gold nanofluids. *Journal of Thermal Analysis and Calorimetry, 114*, pp.557–564. https://doi.org/10.1007/s10973-013-3021-7

Plourde, B.D., Schwalbach, D.B., Abraham, J.P. and Kohler, R.E., 2014. Intracoronary injection of medication from multi-lumen injection catheters. *Journal of Medical Devices, 8*(2), 020902.

Rabbi, H.M.F., Sahin, A.Z., Yilbas, B.S. and Al-Sharafi, A., 2021. Methods for the determination of nanofluid optical properties: A review. *International Journal of Thermophysics, 42*, pp.1–42. https://doi.org/10.1007/s10765-020-02762-0

Rafati, M., Hamidi, A.A. and Niaser, M.S., 2012. Application of nanofluids in computer cooling systems (heat transfer performance of nanofluids). *Applied Thermal Engineering, 45*, pp.9–14. https://doi.org/10.1016/j.applthermaleng.2012.03.028

Sajid, M.U. and Ali, H.M., 2019. Recent advances in application of nanofluids in heat transfer devices: A critical review. *Renewable and Sustainable Energy Reviews, 103*, pp.556–592. https://doi.org/10.1016/j.rser.2018.12.057

Sanjeevi, B. and Loganathan, K., 2020. Synthesis of multi wall carbon nanotubes nanofluid by using two step method. *Thermal Science, 24*(1 Part B), pp.519–524. https://doi.org/10.2298/TSCI190414430S

Septiadi, W.N., Trisnadewi, I.A.N.T., Putra, N. and Setyawan, I., 2018. Synthesis of hybrid nanofluid with two-step method. In *E3S Web of Conferences* (Vol. 67, p. 03057), Kuta, Bali, Indonesia, September 6–8, 2018. EDP Sciences. https://doi.org/10.1051/e3sconf/20186703057.

Sheikhpour, M., Arabi, M., Kasaeian, A., Rokn Rabei, A. and Taherian, Z., 2020. Role of nanofluids in drug delivery and biomedical technology: Methods and applications. *Nanotechnology, Science and Applications, 13*, pp.47–59. https://doi.org/10.2147/NSA.S260374

Shen, M., Chen, H., Zhang, M., Liu, F. and Anh, V., 2022. A comprehensive review of nanofluids with fractional derivatives: Modeling and application. *Nanotechnology Reviews, 11*(1), pp.3235–3249. https://doi.org/10.1515/ntrev-2022-0496

Stark, J.R., Gorman, J.M., Sparrow, E.M., Abraham, J.P. and Kohler, R.E., 2013. *Controlling the rate of penetration of a therapeutic drug into the wall of an artery by means of a pressurized balloon*. https://doi.org/10.4236/jbise.2013.65067

Suresh, S., Venkitaraj, K.P., Selvakumar, P. and Chandrasekar, M., 2011. Synthesis of Al2O3–Cu/water hybrid nanofluids using two step method and its thermo physical properties. *Colloids and Surfaces A: Physicochemical and Engineering Aspects*, *388*(1–3), pp.41–48. https://doi.org/10.1016/j.colsurfa.2011.08.005

Thirunavukarasan, K. and Sucharitha, G., 2025. Thermal performance and MHD peristaltic flow of hybrid nanofluid (Au-Ta/Blood) in an asymmetric conduit with electro-osmosis and shape factor effects. *Case Studies in Thermal Engineering*, *66*, p.105708.

Tielke, J., Maas, M., Castillo, M., Rezwan, K. and Avila, M., 2021. Statistical analysis of thermal conductivity experimentally measured in water-based nanofluids. *Proceedings of the Royal Society A*, 477(2250), p.20210222. https://doi.org/10.1098/rspa.2021.0222

Toghraie, D., Chaharsoghi, V.A. and Afrand, M., 2016. Measurement of thermal conductivity of ZnO–TiO 2/EG hybrid nanofluid: Effects of temperature and nanoparticles concentration. *Journal of Thermal Analysis and Calorimetry*, *125*, pp.527–535. https://doi.org/10.1007/s10973-016-5436-4

Vajravelu, K., Abraham, J.P., Mukhopadhyay, S. and Lakshminarayana, P., 2024. Advances in nanofluid flow, heat, and mass transfer at moving/stretching surfaces. *Advances in Heat Transfer*, *58*, p.81. https://doi.org/10.1016/bs.aiht.2024.05.003

Xuan, Y. and Li, Q., 2000. Heat transfer enhancement of nanofluids. *International Journal of Heat and Fluid Flow*, *21*(1), pp.58–64. https://doi.org/10.1016/S0142-727X(99)00067-3

Zhu, H.T., Lin, Y.S. and Yin, Y.S., 2004. A novel one-step chemical method for preparation of copper nanofluids. *Journal of Colloid and Interface Science*, *277*(1), pp.100–103. https://doi.org/10.1016/j.jcis.2004.04.026

2 Nanofluid Flow at a Moving Surface

2.1 INTRODUCTION

Heat and mass transfer play a crucial role in technological breakthroughs and impact many different research fields. Thermal radiation, a form of heat transfer, has a substantial impact on nanofluid flow and heat transfer properties. Thermal radiation is caused by molecular-level vibrational motion. This thermal motion generates electromagnetic radiation and serves as a route of heat transmission. A short recitation of recent publications in this area will now be provided – these are representative and not intended to be exhaustive. The heat transfer associated with MHD fluid flow under a magnetic field with slip was studied by Alharbi et al. (2023). Heat transfer of MHD Maxwell nanofluid on a vertically moving surface was investigated by Nadeem et al. (2020). Ilango et al. (2024) explored the heat and mass transfer of two non-Newtonian fluids with chemical reactions and thermal radiation. Reddy et al. (2024) explored the heat and mass transmission on a stretching sheet with heat absorption and generation. Idris et al. (2023) explored heat transfer on a permeable moving surface of a magnetized hybrid ferrofluid. Abdel-Wahed et al. (2015) investigated the heat transfer with nonlinear velocity and Brownian motion across a moving surface. Meenakumari et al. (2025) studied the thermal transfer of an MHD Prandtl fluid with heat absorption/generation and suction/injection over a stretching sheet. Other relevant studies include Ajithkumar et al. (2024), Meenakumari et al. (2021), Jahan et al. (2017), Lutera et al. (2024), Samat et al. (2024), and Soomro et al. (2018), which cover transfer under various conditions with similar nanofluid situations.

Heat transfer on a moving surface is an important topic in thermal and fluid mechanics, with the motion of the surface and surrounding fluid both influencing the heat transfer properties and the skin friction. This form of heat transmission is frequently studied in industrial applications such as with conveyor belts, rotating machinery cooling, extrusion processes, and aerodynamic heating on aircraft surfaces, to name a few. Anderson (2010) observed that a boundary layer forms when a fluid's viscosity and adhesion interact with a neighboring surface. Falodun and Fadugba (2017) examined how thermal radiation and viscous dissipation affect the unsteady MHD boundary layer flow of a chemically reactive fluid over a vertical plate. They showed that nanofluids exhibit diverse physical properties and effects on surfaces of varying geometry and arrangement. Bachok et al. (2012) added nanoparticles such as

DOI: 10.1201/9781003571278-2

Al_2O_3, Cu, and TiO_2 to base fluids to assess the separate impact of these materials on the base fluid.

Then, the problem was extended by Najib et al. (2017), who incorporated thermal convective and slip conditions and found a reliable duality of solutions. That study investigated heat transmission and mass diffusion of nanofluids on a moving permeable surface. The surface is constantly diffused, with convective boundary conditions. They also examined the effects of Brownian motion and thermophoresis. Yaseen et al. (2021) studied how heat source/sink, thermal radiation, and velocity slip affect the flow of a hybrid nanofluid (SiO_2-MO_2/*water*) over a moving surface. When a moving surface is subject to convective heat transfer and the interface is porous, a suction/injection condition is formed at the surface. Selected additional studies include Zheng and Zhang (2002), Lin and Huang (1994), Howell et al. (1997), Abedallh et al. (2024), and Ajaykumar et al. (2024), who collectively examined the boundary layer model in relation to nanofluids and other viscous fluids on moving surfaces.

Viscous dissipation is an irreversible procedure by which heat is generated by friction within a fluid or between a fluid and a bounding surface. Viscous dissipation generates a significant temperature increase in polymer manufacturing, as well as high-rate extrusion and injection molding. Swain et al. (2020) investigated dissipating MHD flow submerged in a porous medium incorporating Joule heating. They showed that the Eckert number plays an important role in determining temperature. Gopal et al. (2017) revealed that velocity and viscous dissipation increase in a viscous dissipative Casson nanofluid flow with slip and chemical reactants. Ganga et al. (2015) investigated radiation-driven nanofluid flow across the magnetohydrodynamic boundary layer to an upward plate with a heat source/sink. Goud et al. (2020) examined the impact of thermal radiation and Joule heating on the flow of a Casson nanofluid through a nonlinear inclined porous stretched sheet during chemical reactions. Reddy et al. (2020) studied the impact of heat generation/absorption on MHD copper-water nanofluid movement across a nonlinear stretching/shrinking sheet. Shoaib et al. (2021) explored the effects of thermal energy, concentration, and Brownian motion. The effects of heat-absorption on viscous dissipative and radiation have been investigated in additional previous works but are not referenced here for brevity's sake.

Motivated by the aforementioned references and the wide-ranging potential industrial applications of nanofluids, this study aims to examine the impact nanoparticles on thermal radiation, chemical reactions, and viscous dissipation for flow over a moving surface with a porous medium. Heat generation/absorption and joule heating both influence the temperature and velocity, respectively. The MATLAB® solver bvp5c is utilized to solve the transformed ordinary differential equations numerically. Tables and graphs will be utilized to illustrate the effects of the flow-controlling parameters on the temperature and velocity fields, skin friction coefficient, and heat transfer rate.

2.2 MATHEMATICAL FORMULATION

This example focuses on the two-dimensional steady boundary layer flow of an MHD nanofluid comprising titanium dioxide (TiO_2) nanoparticles over a permeable moving surface. We examine the heat transmission and mass transfer ability of the nanofluid

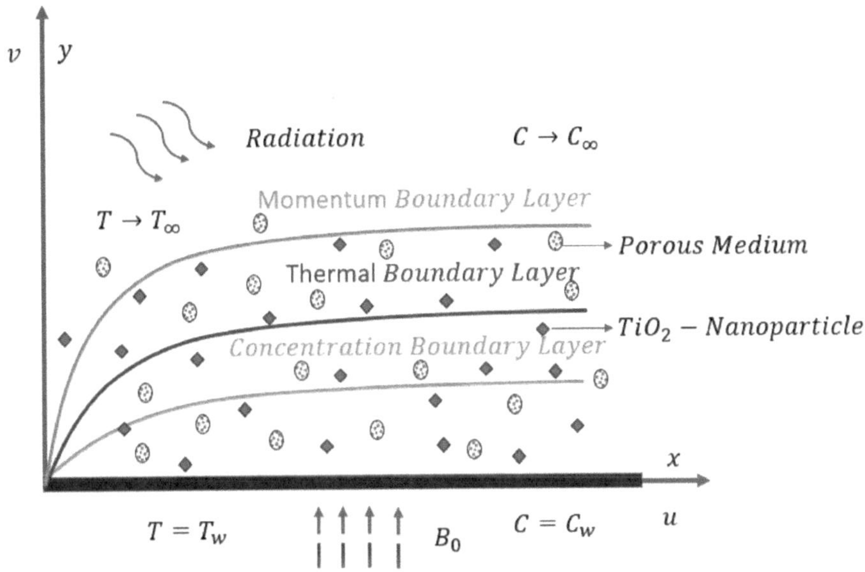

FIGURE 2.1 Sketch of the flow.

flow over a permeable surface, considering thermal radiation, a heat source/sink, and Eckert number along with titanium dioxide (TiO_2) nanoparticles. This investigation involves a moving surface with a porous surface. The magnetic field is oriented perpendicular to the nanofluid flow and the surface. The moving surface is parallel to $x = 0$. The free stream velocity u_∞ and moving surface velocity u_w are constants and they are moving in the same direction. The moving surface maintains a constant temperature and concentration near the surface, T_w and C_w respectively. As $y \to \infty$ the ambient temperature and ambient concentration are T_∞ and C_∞ respectively. The physical geometry that describes the proposed study is shown in Figure 2.1.

The governing equations are provided below.

Mass conservation:

$$\frac{\partial u}{\partial x} + \frac{\partial v}{\partial y} = 0. \tag{2.1}$$

Conservation of momentum:

$$u\frac{\partial u}{\partial x} + v\frac{\partial v}{\partial y} = \frac{\mu_{nf}}{\rho_{nf}}\frac{\partial^2 u}{\partial y^2} - \left(\frac{\sigma}{\rho}\right)_{nf} B_0^2 u - \frac{\mu_{nf}}{\rho_{nf}}\frac{u}{k_p}. \tag{2.2}$$

Conservation of energy:

$$u\frac{\partial T}{\partial x}+v\frac{\partial T}{\partial y}=\left(\frac{k}{\rho C_P}\right)_{nf}\frac{\partial^2 T}{\partial y^2}+\left(\frac{k}{\rho C_P}\right)_{nf}\left(\frac{\partial u}{\partial x}\right)^2$$
$$+\frac{1}{\left(\rho C_P\right)_{nf}}\left(\frac{16\sigma^* T_\infty^3}{3k^*}\right)\frac{\partial^2 T}{\partial y^2}-\frac{1}{\left(\rho C_P\right)_{nf}}Q_0\left(T-T_\infty\right). \tag{2.3}$$

Species conservation:

$$u\frac{\partial C}{\partial x}+v\frac{\partial C}{\partial y}=D_B\frac{\partial^2 C}{\partial y^2}-K_l\left(C-C_\infty\right). \tag{2.4}$$

These governing equations are solved with the boundary conditions (Ajaykumar et al., 2024):

$$\begin{aligned} &u=u_w,\quad v=0,\quad T=T_w,\quad C=C_w \quad at\; y=0\\ &u\to u_\infty,\quad T\to T_\infty,\quad C\to C_\infty \quad as\; y\to\infty. \end{aligned} \tag{2.5}$$

Afzal et al. (1993) and Abdelhafez (1985) established the similarity adjustment on the combined reference velocity $U=u_w+u_\infty$. The largest of the two velocities was adopted as the reference velocity, i.e. $U=\begin{Bmatrix} u_w, & if\;\; u_w>u_\infty \\ u_\infty, & if\;\; u_\infty>u_w \end{Bmatrix}$.

More recent work has generalized the issue to simultaneously moving fluid and surface, albeit for general fluids rather than nanofluids (Abraham and Sparrow, 2005, and Sparrow and Abraham, 2005). The similarity transformations that are used to transform the partial differential equations (2.1–2.4) into ordinary differential equations are provided in Equation (2.6).

$$u=Uf'(\eta),\qquad v=\sqrt{\frac{U\vartheta_f}{2x}}\left(\eta f'(\eta)-f(\eta)\right),$$
$$\eta=y\sqrt{\frac{U}{2x\vartheta_f}},\qquad v_w=-\sqrt{\frac{U\vartheta_f}{2x}}S$$
$$\theta=\frac{\left(T-T_\infty\right)}{\left(T_w-T_\infty\right)},\qquad \chi=\frac{\left(C-C_\infty\right)}{\left(C_w-C_\infty\right)}. \tag{2.6}$$

The transformed equations are:

$$\frac{1}{P_1P_2}f''' + ff'' - \frac{P_3}{P_2}Mf' - \frac{K}{P_1P_2}f' = 0, \tag{2.7}$$

$$\frac{P_6}{P_5}\theta'' + Pr\theta f' + Rd\theta'' + \frac{EcPr}{P_1P_5}(f'')^2 - \frac{Pr}{P_5}Q\theta = 0 \tag{2.8}$$

and

$$\frac{1}{Sc}\chi'' + \chi' f - Kc\chi = 0, \tag{2.9}$$

with transformed boundary conditions:

$$\begin{aligned} &f(0) = 0, \quad f'(0) = 1 - r, \quad \theta(0) = 1, \ \chi(0) = 1 \quad at\ \eta = 0 \\ &f'(\infty) = r, \quad \theta(\infty) = 0, \quad \chi(\infty) = 0 \quad as\ \eta \to \infty. \end{aligned} \tag{2.10}$$

Where $r = \frac{u_\infty}{u_w + u_\infty}$ is the moving parameter with range $(0 \le r \le 1)$ (Afzal et al., 1993).

In this range the moving surface and the free stream are both moving in the same direction. When r is greater than 1, the moving surface travels towards the negative x-direction, and the free stream is moving towards the positive x-direction. If r is less than zero, the moving surface travels towards the positive x-direction, and the free stream travels towards the negative x-direction.

To simply the expressions, multiple dimensionless numbers are formed as shown here:

$$M = \left(\frac{\sigma}{\rho}\right)_f \frac{B_0^2}{U}, \quad K = \frac{k_p U^2}{\vartheta_f^2}, \quad Q = \frac{Q_0}{U(\rho C_p)_f},$$

$$Rd = \frac{4\sigma^* T_\infty^3}{k^* k_f}, \quad Ec = \frac{U^2}{(C_p)_f (T_w - T_\infty)},$$

$$Pr = \frac{(\mu C_p)_f}{k_f}, \qquad Kc = \frac{\vartheta_f K_l (C_w - C_\infty)}{U^2}, \qquad Sc = \frac{\vartheta_f}{D_B}. \tag{2.11}$$

Here, M, K, Q, Rd, Ec, Pr, Kc, Sc are magnetic field, porous parameter, heat source/sink, radiation, Eckert number, Prandtl number, chemical reaction parameter, and Schmidt number, respectively.

2.2.1 Physical Attributes of Ternary Nanofluid and Thermophysical Properties

Various thermophysical properties can be defined for the nanofluid mixture, as shown here for dynamic viscosity, density, specific heat, thermal conductivity, and electrical conductivity.

$$\mu_{nf} = \frac{\mu_f}{\left(1-\phi_3\right)^{2.5}}, \tag{2.12}$$

$$\rho_{nf} = \left(1-\phi_1\right)\rho_f + \phi_1\rho_{S1}, \tag{2.13}$$

$$\left(\rho C_p\right)_{nf} = \left(1-\phi_1\right)\left(\rho C_p\right)_f + \phi_1\left(\rho C_p\right)_{S1}, \tag{2.14}$$

$$k_{nf} = \left(\frac{k_{S1}+2k_f-2\phi_1\left(k_f-k_{S1}\right)}{k_{S1}+2k_f+\phi_1\left(k_f-k_{S1}\right)}\right)k_f, \tag{2.15}$$

$$\sigma_{nf} = \left(\frac{\sigma_{S1}+2\sigma_f-2\phi_1\left(\sigma_f-\sigma_{S1}\right)}{\sigma_{S1}+2\sigma_f+\phi_1\left(\sigma_f-\sigma_{S1}\right)}\right)\sigma_f. \tag{2.16}$$

The following properties ratios will be adopted for simplification purposes:

$$P_1 = \frac{\mu_f}{\mu_{nf}},\quad P_2 = \frac{\rho_{nf}}{\rho_f},\quad P_3 = \frac{\sigma_{nf}}{\sigma_f},\quad P_4 = \frac{k_{nf}}{k_f},\quad P_5 = \frac{\left(\rho C_p\right)_{nf}}{\left(\rho C_p\right)_f}.$$

The thermophysical properties of the titanium oxide nanoparticles and water are given in Table 2.1.

2.2.2 Physical Quantities

The local Nusselt number Nu_x and the coefficient for skin friction (C_f) are the quantities of most significance, and they are defined as follows:

TABLE 2.1
Thermophysical properties for nanoparticles and the base fluid (water)

	Water	TiO_2
$C_p\left(\frac{J}{kg\,K}\right)$	4179	686.2
$\rho\left(\frac{kg}{m^3}\right)$	997.1	4250
$k\left(\frac{W}{m\,K}\right)$	0.613	8.9538
$\sigma\left(\frac{S}{m}\right)$	5.5×10^{-6}	2.6×10^{6}

$$C_f = \frac{\tau_w}{\rho_f U^2},\quad Nu_x = \frac{x\left(q_w + q_r\right)}{k_f\left(T_w - T_\infty\right)} \tag{2.17}$$

where the wall shear stress and heat fluxes are given by:

$$\tau_w = \mu_{nf}\left(\frac{\partial u}{\partial y}\right)_{y=0},\quad q_w + q_r = -\left(k_{nf}\frac{\partial T}{\partial y} + \frac{4\sigma^*}{3k^*}\frac{\partial T^4}{\partial y}\right)_{y=0}. \tag{2.18}$$

Inserting Equation (2.18) and Equation (2.6) into Equation (2.17) yields

$$\left(Re_x\right)^{1/2} C_f = \frac{f''(0)}{P_1\sqrt{2}},\quad \left(Re_x\right)^{-1/2} Nu_x = \frac{-\theta'(0)}{\sqrt{2}}\left(P_6 + \frac{4}{3}Rd\right). \tag{2.19}$$

2.2.3 Numerical Methodology

The numerical solution procedure for the mathematical model is described in this section. Together with an efficient numerical method (bvp5c) functioned by the Lobatto IIIa method and a three-stage collocation algorithm, Equations (2.7), (2.8), (2.9), and their transformed boundary conditions Equation (2.10) are solved numerically by converting the nonlinear ordinary differential equations into linear ordinary differential equations.

The result of this process yields:

$$f''' = P_1P_2\left(-ff'' + \frac{P_3}{P_2}Mf' + \frac{K}{P_1P_2}f\right), \tag{2.20}$$

$$\theta'' = \left(-Pr\theta f' - Rd\theta'' - \frac{EcPr}{P_1P_5}(f'')^2 + \frac{Pr}{P_5}Q\theta\right)\frac{P_5}{P_6} \tag{2.21}$$

and

$$\chi'' = Sc(-\chi' f + Kc\chi), \tag{2.22}$$

with transformed boundary conditions

$$\begin{aligned} &f = 0,\quad f' - 1 + r = 0,\quad \theta = 0 \quad at\ \eta = 0 \\ &f' - r = 0,\quad \theta = 0 \quad as\ \eta \to \infty. \end{aligned} \tag{2.23}$$

The linearization process used to turn equations into a system of linear first-order equations is:

$$\begin{aligned} &f = f(1), f' = f(2), f'' = f(3), \theta = f(4), \theta' = f(5), \\ &\phi = f(6), \phi' = f(7). \end{aligned} \tag{2.24}$$

2.3 RESULTS AND DISCUSSION

This portion of the chapter presents graphical results obtained through the numerical solution of the flow of the nanofluid over the moving surface, as discussed in the preceding section. The velocity, temperature, and concentration profiles are shown for different numerical values of input parameters. In addition, we will provide tabular and graphical representations of variations in skin friction and convective heat transfer coefficients.

Figure 2.2 shows the effect of the magnetic parameter on the velocity profiles. The Lorentz force increases as the magnetic parameter increases and acts in the opposite direction of the flow. Thus, the velocity decreases as the magnetic parameter value increases. An increasing porous parameter decreases the velocity profile, as shown in Figure 2.3. The increase in the porous parameter increases the boundary layer thickness, and the velocity of the nanofluid decreases. Figure 2.4 depicts the effect of the magnetic parameter on the temperature profile. As M grows, the temperature profile diminishes. This is because the magnetic field creates a Lorentz force that opposes fluid motion, resulting in higher viscous dissipation and decreased heat transfer. The thermal boundary layer's thickness reduces as M increases. This is because the magnetic field causes greater convective cooling.

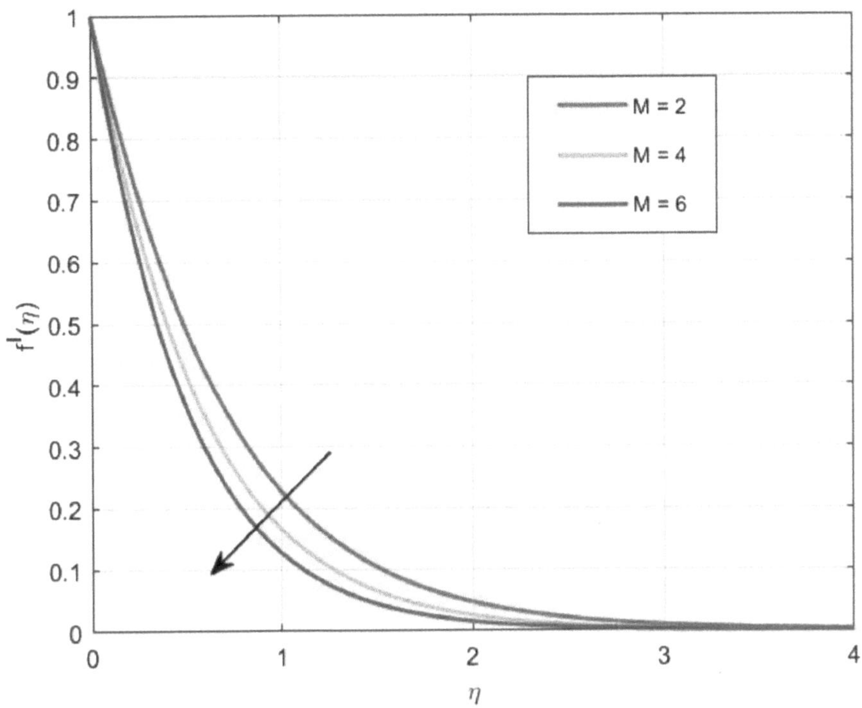

FIGURE 2.2 Velocity profile for the variations of the magnetic parameter.

Figure 2.5 shows how temperature profiles change as the heat source/sink parameter varies. When the heat source parameter Q is positive, heat is generated within the system during flow; when Q is negative, heat is absorbed within the system. As Q becomes more negative, the temperature decreases.

Figure 2.6 shows how the temperature increases as the radiation parameter increases. When radiation predominates over conduction during the nanofluid flow, the temperature rises as the radiation parameter increases as a result of the enhanced heat transfer in the boundary layer region. Figure 2.7 illustrates how the temperature rises in tandem with the Eckert number value. The conversion of kinetic energy into thermal energy is enhanced and the temperature of the nanofluid flow increases by increasing the Eckert number.

Figure 2.8 is plotted to show how the porous parameter affects the temperature profile. The temperature of the fluid is raised as the porous parameter increases. In general, higher porosity impacts reducing the resistance of fluid flow. Heat transfer is enhanced, and fluid circulation inside the porous media is made possible. The medium's internal temperature tends to rise. The impact of chemical reaction on the concentration profile is depicted in Figure 2.9. The observation shows that the concentration of the nanofluid decreases as we raise the chemical reaction parameter. The interaction of the nanoparticles with the chemical reaction weakens the concentration boundary layer and decreases the concentration of the nanofluid.

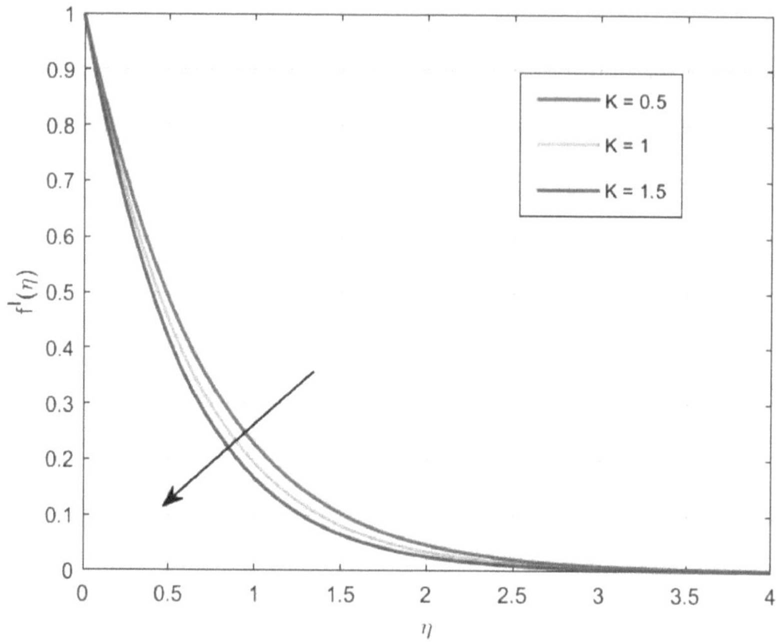

FIGURE 2.3 Velocity profile for the variations of porosity parameter.

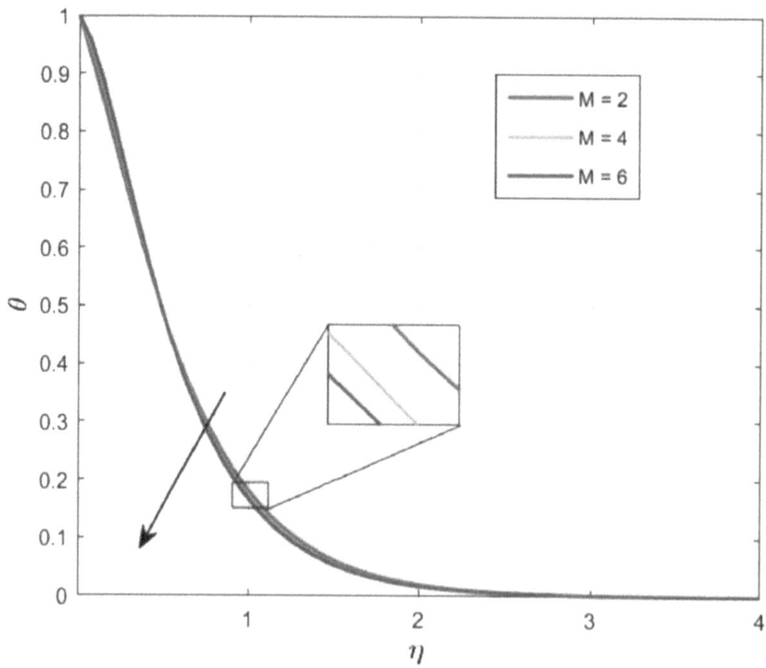

FIGURE 2.4 Temperature profile for the variations of the magnetic parameter.

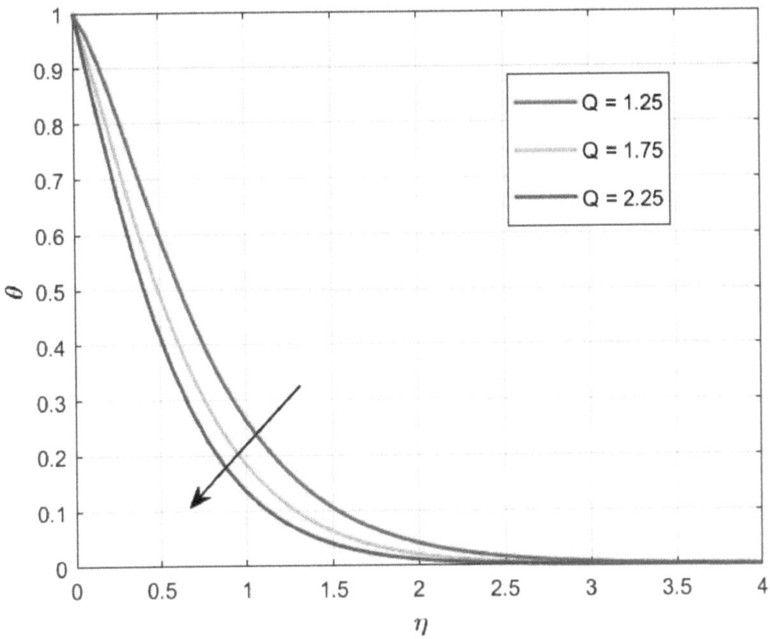

FIGURE 2.5 Temperature profile for variations of the heat source/sink parameter.

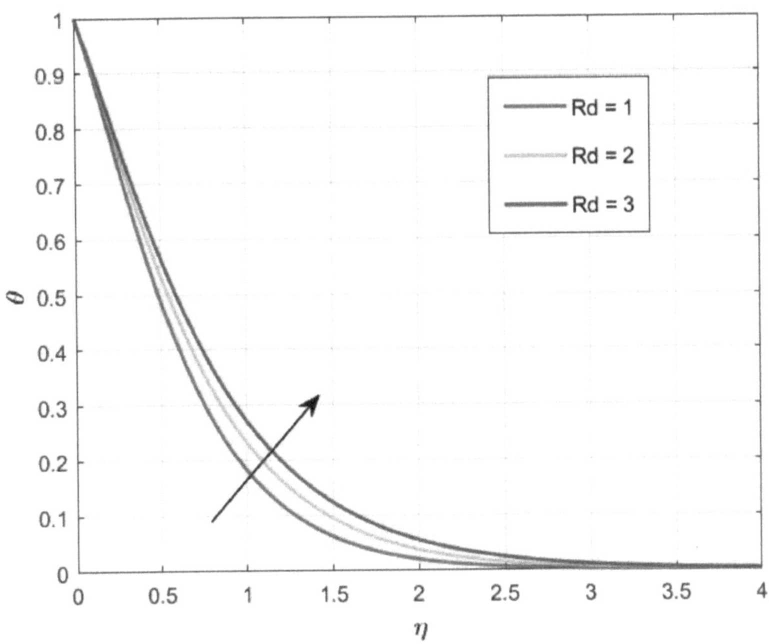

FIGURE 2.6 Temperature profile for variations of the radiation parameter.

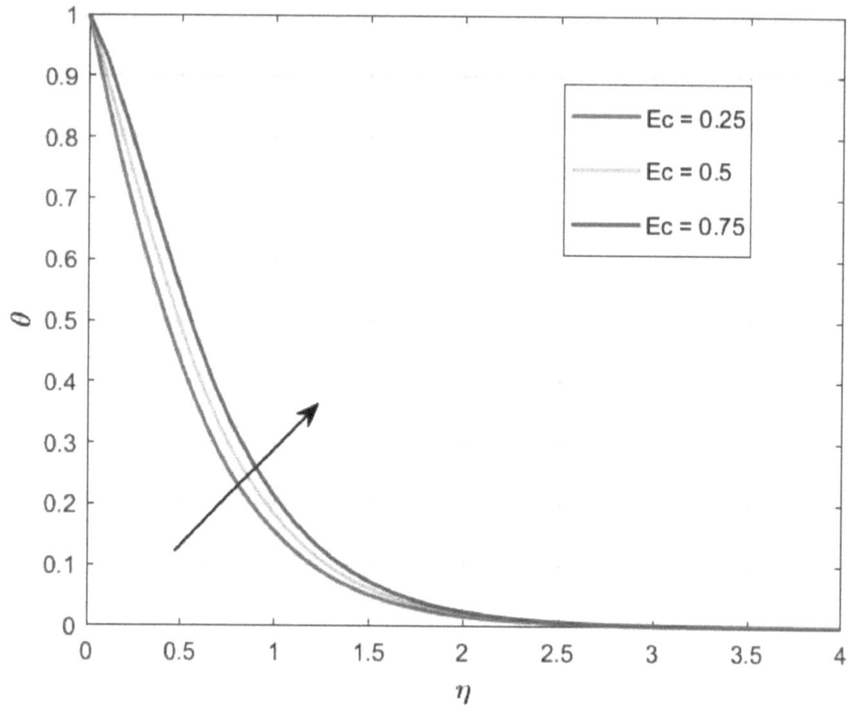

FIGURE 2.7 Temperature profile for variations of the Eckert number.

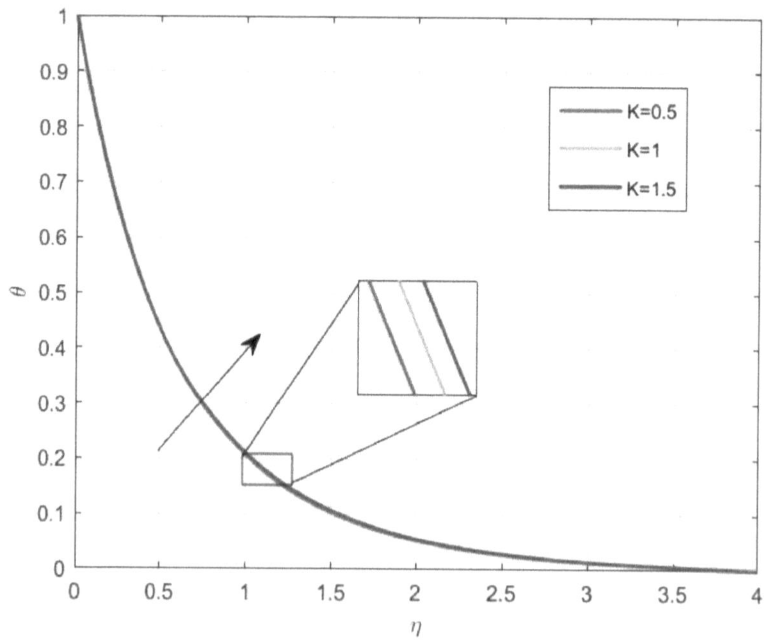

FIGURE 2.8 Temperature profile for variations of the porosity parameter.

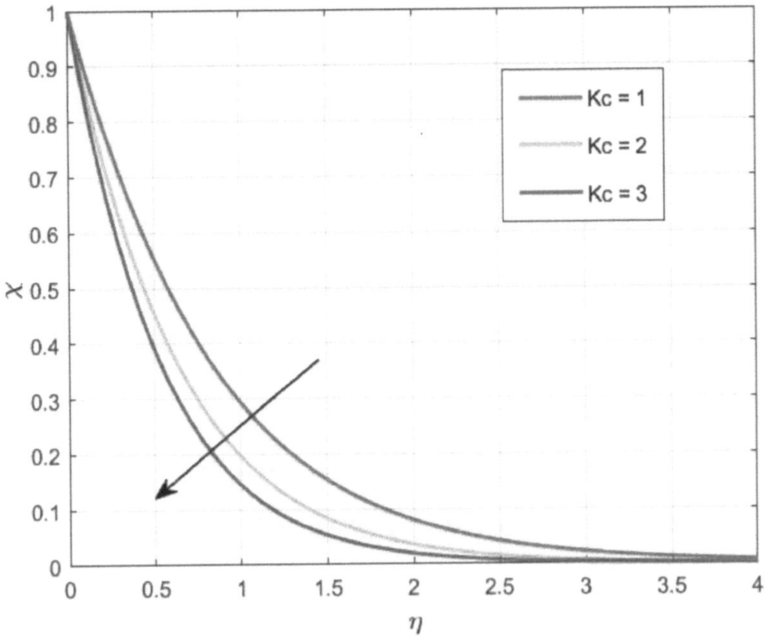

FIGURE 2.9 Concentration profile for variations of the chemical reaction parameter.

Figure 2.10 demonstrates that the concentration increases with increasing magnetic field. The variations of the concentration profile with the Schmidt number are depicted in Figure 2.11. The concentration decreases for raising the values of the Schmidt number.

Figure 2.12 displays the variations of magnetic field on skin friction by assuming K is a constant. The impact of porosity on skin friction is presented in Figure 2.13. The impacts of the radiation parameter and the Eckert number are displayed in Figures 2.14 and 2.15, respectively.

The numerical values for skin friction and the Nusselt number are given in Table 2.2. These results explain that skin friction becomes more negative for increasing values of the magnetic parameter and the porous parameter. The Nusselt number is enhanced with increasing thermal radiation parameters.

2.4 CONCLUDING REMARKS

The two-dimensional steady boundary layer flow of an MHD nanofluid containing titanium dioxide nanoparticles across a permeable moving surface in the presence of viscous dissipation and heat source/sink is investigated in this study. A numerical method (MATLAB® bvp5c) was used to solve the ordinary differential equations that were well obtained from similarity transformations of the governing equations. The main findings of this study are listed below.

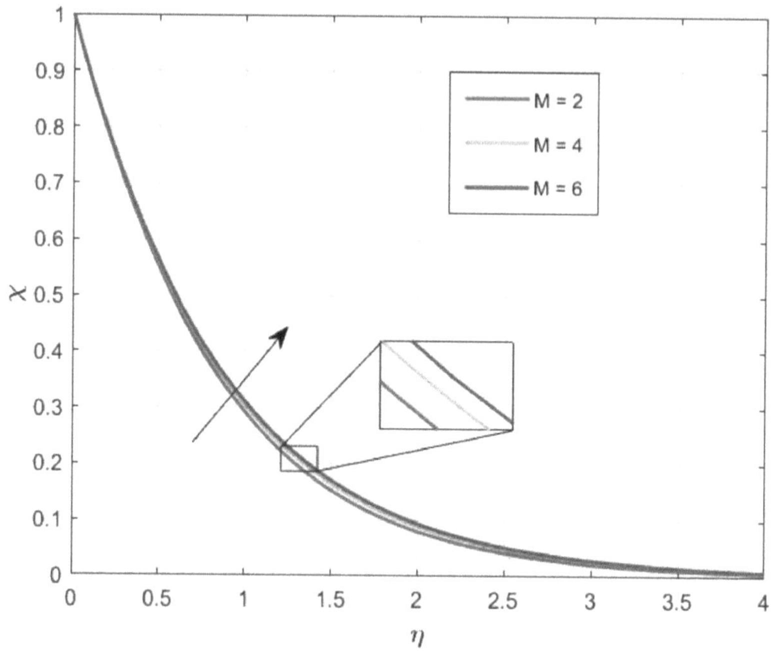

FIGURE 2.10 Concentration profile for the variations of the magnetic parameter.

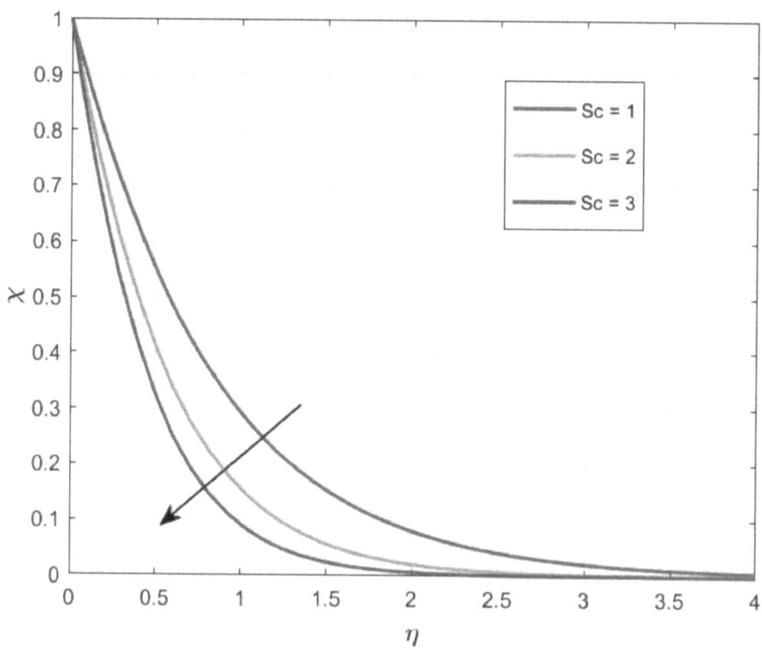

FIGURE 2.11 Concentration profile for the variations of Schmidt number.

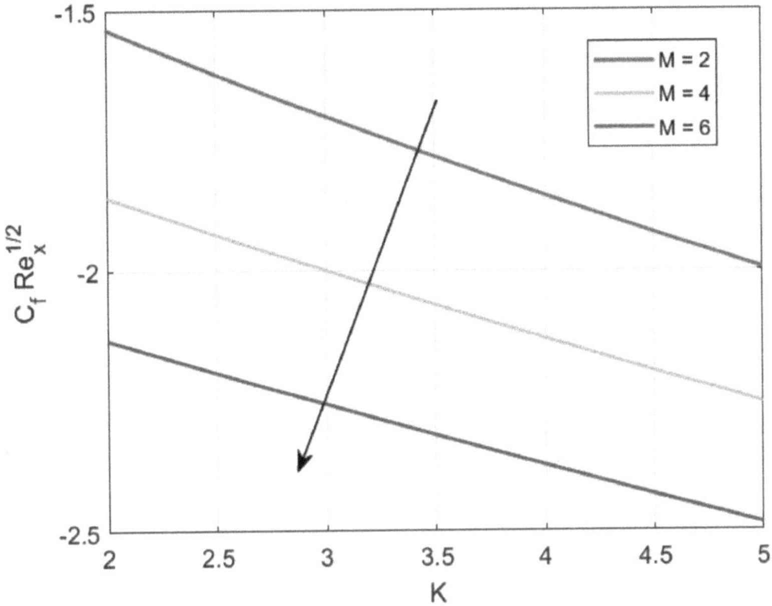

FIGURE 2.12 Skin friction for the variations of the magnetic parameter.

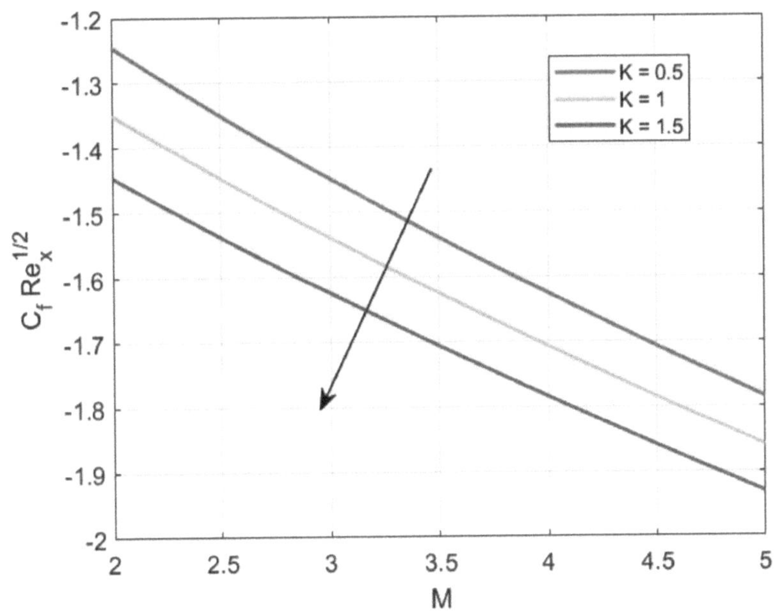

FIGURE 2.13 Skin friction for the variations of the porosity parameter.

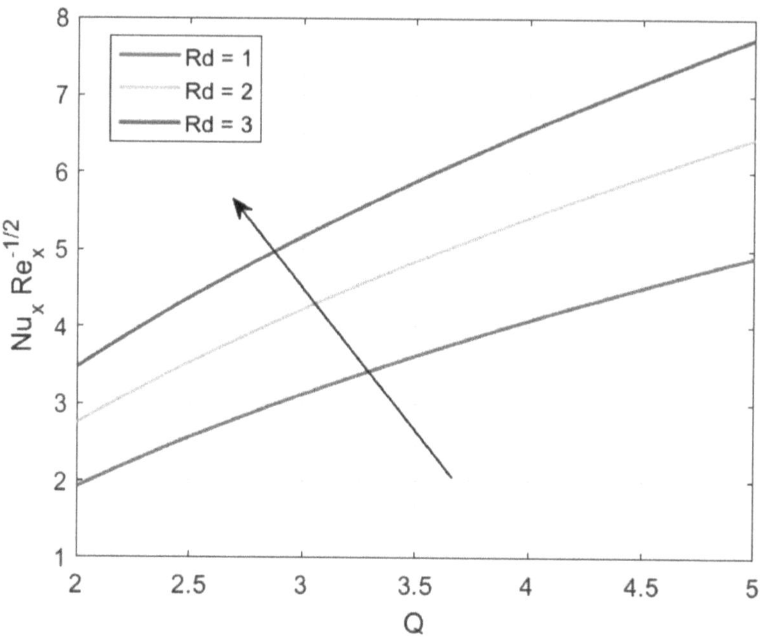

FIGURE 2.14 Nusselt number for the variations of the radiation parameter.

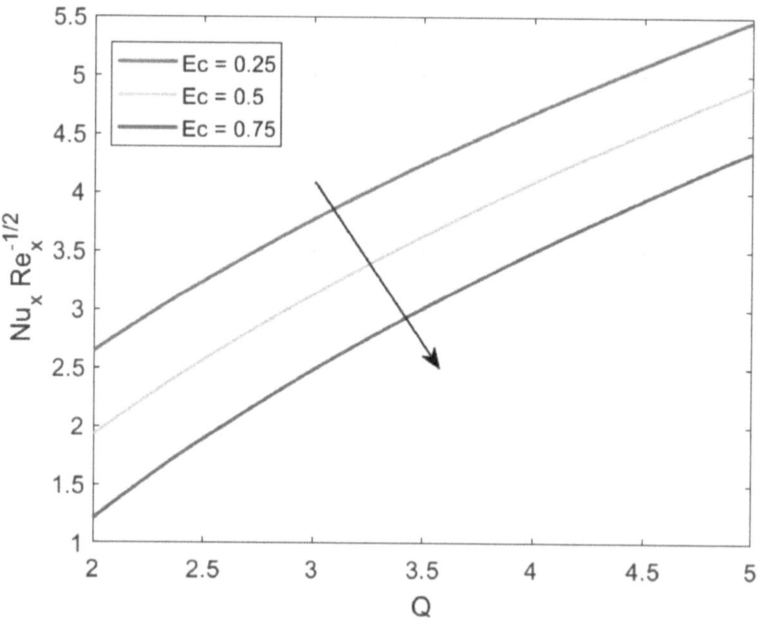

FIGURE 2.15 Nusselt number for the variations of the Eckert number.

TABLE 2.2
Numerical values for skin friction and the Nusselt number

M	K	Q	Rd	Ec	$C_f Re_x^{\frac{1}{2}}$	$Nu_x Re_x^{\frac{-1}{2}}$
2					-1.24558	1.56377
4					-1.62583	1.05152
6					-1.93298	0.6112
	0.5				-1.24558	1.56377
	1				-1.34973	1.4279
	1.5				-1.44647	1.29849
		1.25				0.71999
		1.75				1.56377
		2.25				2.25514
			1			1.56377
			2			2.30986
			3			2.97329
				0.25		2.30135
				0.5		1.56377
				0.75		0.82619

1. The increase in magnetic field and porous parameter decreases the velocity of the fluid.
2. As the Eckert number increases, the transportation of the temperature is boosted, and the temperature profile of the nanofluid is increased.
3. The enhance in radiation parameter increases the temperature of nanofluid.
4. The concentration of nanoparticles is reduced by higher Schmidt number and chemical reaction parameter values.

REFERENCES

Abdelhafez, T.A., 1985. Skin friction and heat transfer on a continuous flat surface moving in a parallel free stream. *International Journal of Heat and Mass Transfer*, *28*(6), pp.1234–1237.

Abedallh, A.S., Alomar, O.R. and Yasin, N.J., 2024. Numerical and experimental investigation on mixed convection heat transfer inside cavity heated from below with reciprocating moving upper surface. *International Communications in Heat and Mass Transfer*, *159*, p.108242.

Abdel-Wahed, M.S., Elbashbeshy, E.M.A. and Emam, T.G., 2015. Flow and heat transfer over a moving surface with nonlinear velocity and variable thickness in a nanofluids in the presence of Brownian motion. *Applied Mathematics and Computation*, *254*, pp.49–62.

Abraham, J.P., and Sparrow, E.M., 2005. Friction drag resulting from the simultaneous imposed motion of a freestream and its bounding surface. *International Journal of Heat and Fluid Flow*, 26, pp.289–295.

Afzal, N., Badaruddin, A. and Elgarvi, A.A., 1993. Momentum and heat transport on a continuous flat surface moving in a parallel stream. *International Journal of Heat and Mass Transfer*, *36*(13), pp.3399–3403.

Ajaykumar, M., Ajay, C.K. and Srinivasa, A.H., 2024. Effects of viscous dissipation, internal heat source/sink and Prandtl number on flow and heat transfer in a moving fluid over a moving flat surface with an applied magnetic field. *Materials Today: Proceedings*, *100*, pp.4–8.

Ajithkumar, M., Lakshminarayana, P., Sucharitha, G. and Vajravelu, K., 2024. Investigation of convective peristaltic flow of non-Newtonian fluids through a non-uniform tapered porous conduit with Ohmic heating and viscous dissipation. *International Journal of Modelling and Simulation*, (in press). doi: 10.1080/02286203.2024.2345259.

Alharbi, K.A.M., Ullah, Z., Jabeen, N. and Ashraf, M., 2023. Magnetohydrodynamic and thermal performance of electrically conducting fluid along the symmetrical and vertical magnetic plate with thermal slip and velocity slip effects. *Symmetry*, *15*(6), p.1148.

Anderson, J., 2010. *Fundamentals of aerodynamics*, 4th ed. New York: McGraw Hill.

Bachok, N., Ishak, A. and Pop, I., 2012. Flow and heat transfer characteristics on a moving plate in a nanofluid. *International Journal of Heat and Mass Transfer*, *55*(4), pp.642–648.

Falodun, B.O. and Fadugba, S.E., 2017. Effects of heat transfer on unsteady magnetohydrodynamics (MHD) boundary layer flow of an incompressible fluid on a moving vertical plate. *World Scientific News*, *88*(2), pp.118–137.

Ganga, B., Ansari, S.M.Y., Ganesh, N.V. and Hakeem, A.A., 2015. MHD radiative boundary layer flow of nanofluid past a vertical plate with internal heat generation/absorption, viscous and ohmic dissipation effects. *Journal of the Nigerian Mathematical Society*, *34*(2), pp.181–194.

Gopal, D., Kishan, N. and Raju, C.S.K., 2017. Viscous and Joule's dissipation on Casson fluid over a chemically reacting stretching sheet with inclined magnetic field and multiple slips. *Informatics in Medicine Unlocked*, *9*, pp.154–160.

Goud, B.S., Reddy, Y.D. and Rao, V.S., 2020. Thermal radiation and Joule heating effects on a magnetohydrodynamic Casson nanofluid flow in the presence of chemical reaction through a nonlinear inclined porous stretching sheet. *Journal of Naval Architecture and Marine Engineering*, *17*(2), pp.143–164.

Howell, T.G., Jeng, D.R. and De Witt, K.J., 1997. Momentum and heat transfer on a continuously moving surface in a power law fluid. *International Journal of Heat and Mass Transfer*, *40*(8), pp.1853–1861.

Idris, S., Jamaludin, A., Nazar, R. and Pop, I., 2023. Heat transfer characteristics of magnetized hybrid ferrofluid flow over a permeable moving surface with viscous dissipation effect. *Heliyon*, *9*(5), e159907.

Ilango, M.S., Lakshminarayana, P., Sucharitha, G., Vinodkumar Reddy, M. and Vajravelu, K., 2024. Analysis of entropy optimization in MHD flow of non-newtonian nanofluids with chemical reaction and thermal energies. *Journal of Computational and Theoretical Transport*, *54*, pp. 68–93.

Jahan, S., Sakidin, H., Nazar, R. and Pop, I., 2017. Boundary layer flow of nanofluid over a moving surface in a flowing fluid using a revised model with stability analysis. *International Journal of Mechanical Sciences*, *131*, pp.1073–1081.

Lin, H.T. and Huang, S.F., 1994. Flow and heat transfer of plane surfaces moving in parallel and reversely to the free stream. *International Journal of Heat and Mass Transfer*, *37*(2), pp.333–336.

Lutera, J.N., Shekar, M.R. and Goud, B.S., 2024. Analyzing the exothermic reactions of Cu nanoparticles with the Dufour effect and the influence of the heat generation and permeability past a porous moving surface. *Results in Chemistry*, *10*, p.101730.

Meenakumari, R., Lakshminarayana, P. and Vajravelu, K., 2021. Unsteady MHD flow of a Williamson nanofluid on a permeable stretching surface with radiation and chemical reaction effects. *The European Physical Journal Special Topics*, *230*, pp.1355–1370.

Meenakumari, R., Yamuna Rani, P., Sucharitha, G., Vajravel, K. and Lakshminarayana, P., 2025. Thermal inspection on MHD Prandtl fluid flow past a stretching sheet in the presence of heat generation/absorption and suction/injection. *Physica Scripta*, *100*, article no. 025237.

Nadeem, S., Akhtar, S. and Abbas, N., 2020. Heat transfer of Maxwell base fluid flow of nanomaterial with MHD over a vertical moving surface. *Alexandria Engineering Journal*, *59*(3), pp.1847–1856.

Najib, N., Bachok, N., Arifin, N.M. and Senu, N., 2017. Boundary layer flow and heat transfer of nanofluids over a moving plate with partial slip and thermal convective boundary condition: Stability analysis. *International Journal of Mechanics*, *11*(1), pp.18–24.

Reddy, M.V., Meenakumari, R., Sucharitha, G., Ali, F., Zafar, S.S., and Lakshminarayana, P., 2024. Heat and mass transfer analysis of conducting non-Newtonian nanofluid flows over an elongating sheet with a non-uniform heat source. *Modern Physics Letters B*, 38, p.2450349.

Reddy, Y.D., Rao, V.S. and Kumar, M.A., 2020, July. Effect of heat generation/absorption on MHD copper-water nanofluid flow over a nonlinear stretching/shrinking sheet. In *AIP Conference Proceedings* (Vol. 2246, No. 1). Hyderabad, India: AIP Publishing.

Samat, N.A.A., Bachok, N. and Arifin, N.M., 2024. Magnetohydrodynamic flow of carbon nanotubes and heat transfer over a moving thin Needle: A numerical and research surface methodology. *Ain Shams Engineering Journal*, *15*, p.102833.

Shoaib, M., Khan, R.A., Ullah, H., Nisar, K.S., Raja, M.A.Z., Islam, S., Felemban, B.F. and Yahia, I.S., 2021. Heat transfer impacts on Maxwell nanofluid flow over a vertical moving surface with MHD using stochastic numerical technique via artificial neural networks. *Coatings*, *11*(12), p.1483.

Soomro, F.A., Usman, M., Haq, R.U. and Wang, W., 2018. Melting heat transfer analysis of Sisko fluid over a moving surface with nonlinear thermal radiation via collocation method. *International Journal of Heat and Mass Transfer*, *126*, pp.1034–1042.

Sparrow, E.M. and Abraham, J.P., 2005. Universal solutions for the streamwise variation of the temperature of a moving sheet in the presence of a moving fluid. *International Journal of Heat and Mass Transfer*, 48, pp.3047–3056.

Swain, B.K., Parida, B.C., Kar, S. and Senapati, N., 2020. Viscous dissipation and joule heating effect on MHD flow and heat transfer past a stretching sheet embedded in a porous medium. *Heliyon*, *6*(10), p. e05338.

Yaseen, M., Kumar, M. and Rawat, S.K., 2021. Assisting and opposing flow of a MHD hybrid nanofluid flow past a permeable moving surface with heat source/sink and thermal radiation. *Partial Differential Equations in Applied Mathematics*, *4*, p.100168.

Zheng, L.C. and Zhang, X.X., 2002. Skin friction and heat transfer in power-law fluid laminar boundary layer along a moving surface. *International Journal of Heat and Mass Transfer*, *45*(13), pp.2667–2672.

3 Nanofluid Flow, Heat, and Mass Transfer at a Stretching Sheet

3.1 NANOFLUID FLOW PAST A VERTICAL LINEAR STRETCHING SHEET

Fluid flow, heat, and mass transfer past a linear stretching sheet have received a lot of attention in the research community for their various applications in many engineering disciplines. In many manufacturing processes, a stretching surface mechanically and thermally interacts with the ambient fluid as discussed in Pal (2009). The quality of a final product often depends on both flow and heat transfer processes during fabrication (see, for example, Mukhopadhyay et al., 2012). Multiple studies have explored the influence of operating conditions on thermal fluid processing, including Zeeshan and Majeed (2016), Salahuddin et al. (2016), Majeed et al. (2017), and Ahmad and Ishak (2017). Hussain et al. (2020) highlighted the features of visco-inelastic fluid flow caused by a stretching sheet in the presence of an external magnetic field. They used a generalized Fourier heat conduction law in their study. Turner and Weidman (2021) focused on stagnation point flow on a biaxially stretching surface. Loganathan et al. (2021) analyzed Maxwell fluid flow due to a radiative stretching surface. They reported the combined effects of a magnetic field and Cattaneo–Christov dual diffusion on flow and heat transfer characteristics. These are recent, representative studies in this discipline.

A large body of recent research has focused on linear stretching, including Pal and Mandal (2014), Haroun (2017), Sandeep and Sulochana (2018), Loganathan (2020a), Loganathan (2020b), Elanchezhian (2020), and Khana et al. (2020). Using a Cattaneo–Christov double-diffusion model, Negi et al. (2021) reported the effects of Stefan blowing on Cu-water nanofluid flow past a stretching surface. Using a two-phase model for the nanofluid, Sulochana and Mahalaxmi (2022) explored the combined effects of thermal radiation and chemical reaction on a Williamson nanofluid flow past a stretching surface. Mandal and Pal (2023) estimated the entropy generation for cross-hybrid nanofluid flow past a stretching sheet embedded in a Darcy-Forchheimer porous medium in the presence of a magnetic field and quadratic thermal radiation. Renuka et al. (2023) explored the combined effects of thermal radiation, viscous dissipation, magnetic fields, and slip on non-Newtonian nanofluid flow over a stretching surface. They utilized a Casson fluid model to present the non-Newtonian behavior of the fluid. With a second-grade fluid model, Dey and Borah

DOI: 10.1201/9781003571278-3

(2023) explored nanofluid flow due to a stretching sheet in the presence of an externally applied magnetic field and analyzed the convective energy transfer. Rashid et al. (2024) inspected the micropolar nanofluid flow caused by a linear stretching surface with Cu as the nanoparticle material and water as the base fluid. They reported the effects of the shape of the nanoparticles on the thermal-fluid results. Mahmood et al. (2024) presented a comparative analysis of a hybrid nanofluid flow over a stretching sheet near a stagnation point. They discussed the combined effects of the magnetic field, viscous dissipation, and suction. Gamar et al. (in press) highlighted the effects of thermal radiation on MHD micropolar nanofluid flow over a stretching surface in the presence of Navier slip near a stagnation point.

Motivated by these research studies, we are interested in presenting nanofluid flow past a vertical, linearly stretching surface using a two-phase nanofluid model.

3.1.1 Mathematical Analysis of the Problem

Let (u, v) be the velocity components in the x and y directions, respectively. The continuity, momentum, and energy equations for steady two-dimensional flow of a nanofluid past a heated vertical linearly stretching sheet (see Figure 3.1) can be written as

$$\frac{\partial u}{\partial x} + \frac{\partial v}{\partial y} = 0, \tag{3.1}$$

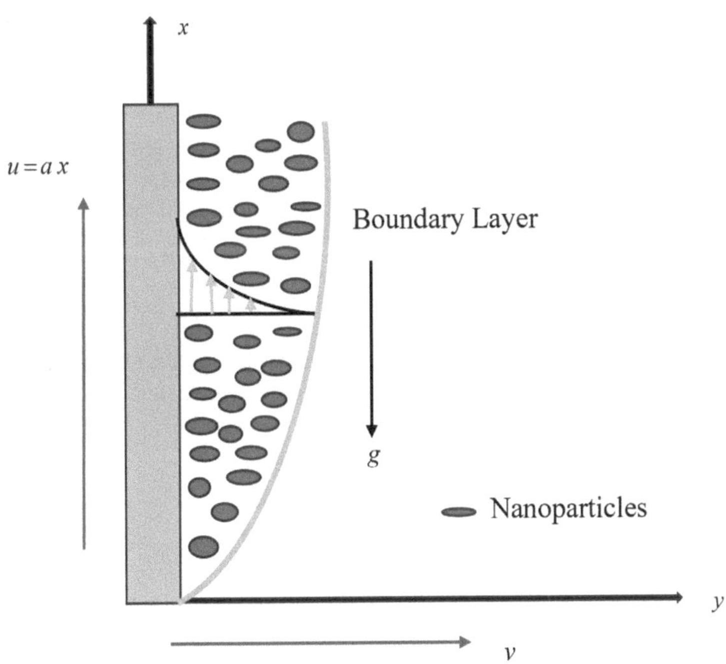

FIGURE 3.1 Geometry of the physical flow problem.

$$u\frac{\partial u}{\partial x}+v\frac{\partial u}{\partial y}=\nu\frac{\partial^2 u}{\partial y^2}+g\beta(T-T_\infty), \tag{3.2}$$

$$u\frac{\partial T}{\partial x}+v\frac{\partial T}{\partial y}=\frac{\kappa}{\rho c_p}\frac{\partial^2 T}{\partial y^2}+\frac{\left(\rho c_p\right)_p}{\left(\rho c_p\right)_f}\left[D_B\frac{\partial C}{\partial y}\frac{\partial T}{\partial y}+\frac{D_T}{T_\infty}\left(\frac{\partial T}{\partial y}\right)^2\right], \tag{3.3}$$

$$u\frac{\partial C}{\partial x}+v\frac{\partial C}{\partial y}=D_B\frac{\partial^2 C}{\partial y^2}+\frac{D_T}{T_\infty}\frac{\partial^2 T}{\partial y^2}. \tag{3.4}$$

Here, T is the temperature, κ is the thermal conductivity, ρ is the fluid density, ν is the kinematic viscosity of the fluid, C is the concentration of the nanomaterial within the fluid, $(\rho c_p)_p$ and $(\rho c_p)_f$ represent specific heat capacities of the nanoparticles and the base fluid, respectively. D_B is the coefficient of Brownian diffusion, D_T represents the coefficient of thermophoresis diffusion where $D_T = D_T'/x$, and D_T' is the initial value of D_T.

The boundary conditions are:

$$u = ax,\ v = 0,\ T = T_w(x) = T_\infty + b_1 x, C = C_w \text{ at } y = 0, \tag{3.5}$$

$$u \to 0, T \to T_\infty, C \to C_\infty \text{ as } y \to \infty \tag{3.6}$$

where $a > 0$ is the stretching rate of the sheet, T_w is the variable wall temperature, b_1 is a constant, and T_∞ and C_∞ are the free stream temperature and concentration, respectively.

The following similarity transformations are introduced

$$u = axf'(\eta),\ v = -\sqrt{a\nu}f(\eta), \eta = \sqrt{\frac{a}{\nu}}y,\ \theta = \frac{T-T_\infty}{T_w-T_\infty}, \phi = \frac{C-C_\infty}{C_w-C_\infty}, \tag{3.7}$$

where θ is the dimensionless temperature, φ is the dimensionless concentration, and η is the similarity variable. Equation (3.1) is automatically satisfied by (3.7). Inserting Equation into Equations (3.2)–(3.4) yields:

$$f''' + ff'' - f'^2 + \lambda\theta = 0, \tag{3.8}$$

$$\frac{1}{Pr}\theta'' + f\theta' + Nb\,\theta'\phi' + Nt\,\theta'^2 = 0, \tag{3.9}$$

$$\phi'' + Le\,f\phi' + \frac{Nt}{Nb}\theta'' = 0 \tag{3.10}$$

Multiple dimensionless numbers involved here are defined as follows:

$$\lambda = g\beta b_f/a^2,$$
$$Nb = (\rho c_p)_p D_B (C_w - C_\infty)/(\rho c_p)_f v,$$
$$Nt = (\rho c_p)_p D_T' b_f/(\rho c_p)_f T_\infty v,$$
$$Le = v/D_B,$$
$$Pr = v\rho c_p/\kappa,$$

where λ is the mixed convection parameter.

The boundary conditions now become:

$$f'(\eta) = 1, f(\eta) = 0, \theta(\eta) = 1, \phi(\eta) = 1 \text{ at } \eta = 0 \tag{3.11}$$

$$f'(\eta) \to 0, \theta(\eta) \to 0, \phi(\eta) \to 0 \text{ as } \eta \to \infty. \tag{3.12}$$

3.1.2 Numerical Results and Discussion

The numerical solutions to the above boundary value problem, set forth in Equations (3.8)–(3.12), are solved as an initial value problem using the Runge-Kutta method. The numerical results are obtained for the region $0 < \eta < \eta_\infty$, where $\eta_\infty = 7$ corresponds to the edge of the boundary layers.

For the verification of our numerical code, the temperature gradients at the surface are compared with available literature data and presented in Table 3.1. The data shows a perfect agreement with the results of Ishfaq et al. (2016), Khan and Pop (2010), Wang (1989), and Kandasamy et al. (2011).

First, we concentrate on the effects of the mixed convection parameter λ on velocity, temperature, and concentration. From Figure 3.2a, it is seen that the velocity field is found to decay with the increasing value of η for all the values of λ considered. Additionally, fluid velocity is found to increase with the increasing values of λ near the middle of the boundary layer, yet the temperature (Figure 3.2b) and concentration (Figure 3.2c) distributions express a different behavior.

Fluid velocity diminishes with the rise in the Prandtl number for positive values of the mixed convection parameter λ, i.e., for buoyancy-aided flow. An opposite nature of velocity is noted for negative values of λ, i.e., for buoyancy-opposed flow (see

TABLE 3.1
Values of $-\theta'(0)$ for various values of the Prandtl number, *Pr*

Pr	Ishfaq et al. (2016)	Khan and Pop (2010)	Wang (1989)	Kandasamy et al. (2011)	Present results
0.7	0.4539		0.4539	0.4542	0.4539
2	0.9114	0.9113	0.9114	0.9114	0.9114
7	1.8954	1.8954	1.8954	1.8952	1.8954
20	3.3539	3.3539	3.3539	-	3.3539

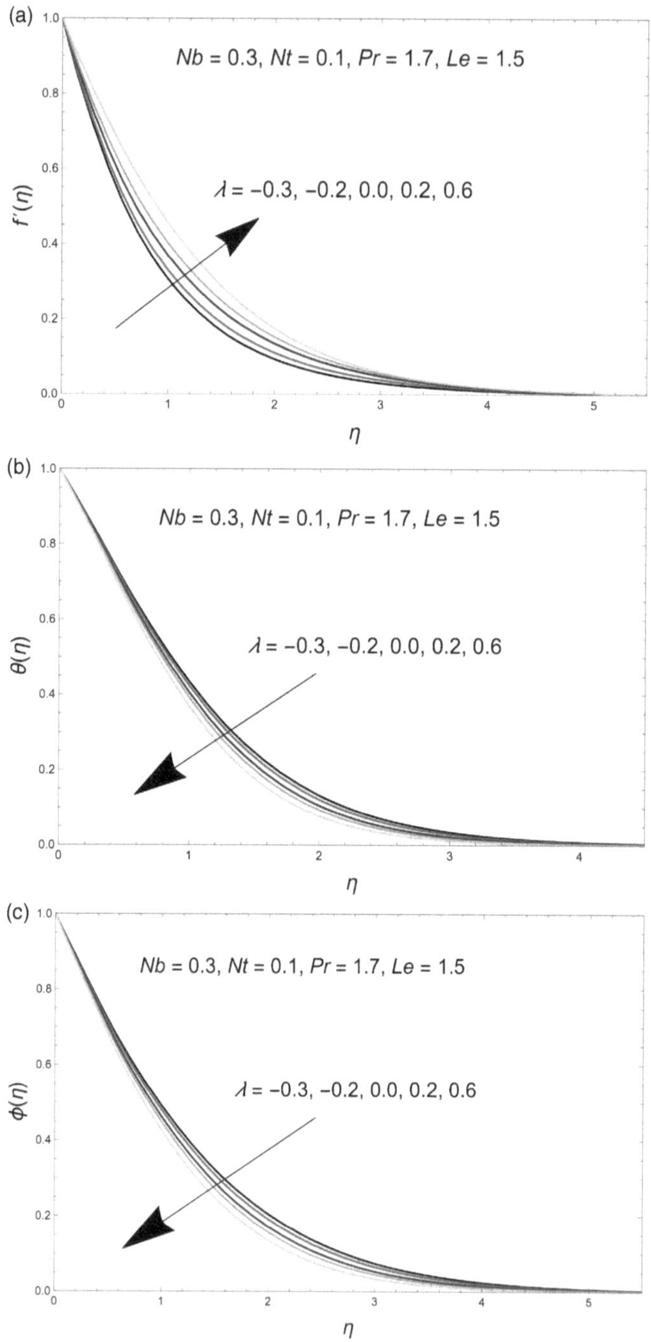

FIGURE 3.2 (a) Velocity, (b) temperature, and (c) concentration profiles for several values of the buoyancy parameter λ.

Figure 3.3a). But the temperature is reduced for rising values of Prandtl number for both buoyancy-aided and opposed flows as shown in Figure 3.3b. Though near the surface, concentration slightly rises with the rise in Prandtl number, but later, the concentration diminishes for an increase in *Pr* (see Figure 3.3c). This effect is not very significant.

The concentration diminishes with the rise in Lewis number, *Le*, for both cases of buoyancy-aided and opposed flows (see Figure 3.4).

Fluid velocity rises slightly with the rise in Brownian motion parameter, *Nb*, for buoyancy-aided flow, but the opposite trend of velocity is noted for buoyancy-opposed flow (Figure 3.5a). The temperature of the nanofluid rises significantly for the rise in Brownian motion parameter, *Nb* (see Figure 3.5b). However the concentration is found to reduce with rising *Nb* for all values of λ, as displayed in Figure 3.5c.

Temperature rises with the rise in thermophoresis parameter, *Nt*, for both the cases of buoyancy-aided and opposed flows (see Figure 3.6a). However, the effect is not significant. On the other hand, significant changes in concentration are noted for growing values of *Nt*. The nanoparticle concentration rises rapidly with the rise in thermophoresis parameter, *Nt*, for both the cases of buoyancy-aided and opposed flows, as evident from Figure 3.6b.

The absolute value of surface shear stress diminishes with rising mixed convection parameters (Table 3.2). The absolute value of temperature gradient and concentration gradient at the surface, i.e., $\theta'(\mathbf{0})$ and $\varphi'(\mathbf{0})$, both rise with rise in λ (Table 3.2). Negative values of $\theta'(\mathbf{0})$ and $\varphi'(\mathbf{0})$ physically indicate that heat and mass are transferred from the surface to the fluid. Thus, a significant cooling effect on the surface is obtained.

Absolute values of surface shear stress and temperature gradient at the surface are both found to rise with rising values of *Pr* (Table 3.3). Still, the absolute value of the concentration gradient at the surface is reduced for larger values of *Pr* (Table 3.3).

Absolute values of surface shear stress and temperature gradient at the surface both diminish with a rise in Lewis number (Table 3.4). Still, the absolute value of the concentration gradient at the surface rises with *Le* (Table 3.4).

With a rise of *Nb*, absolute values of surface shear stress and temperature gradient at the surface both diminish (Table 3.5). Still, the absolute values of the concentration gradient at the surface rise rapidly with *Nb* (Table 3.5).

With a rise in *Nt*, absolute values of surface shear stress, wall temperature gradient, and concentration gradient at the surface diminish, as shown in Table 3.6. These observations agree with the earlier findings.

3.2 NANOFLUID FLOW PAST A NONLINEARLY STRETCHING SHEET

Most of the available literature on this topic deals with nanofluid flow due to a linear stretching surface. In such cases, the velocity of the surface is linearly proportional to the origin (fixed), as discussed in Vajravelu and Mukhopadhyay (2015). However, Gupta and Gupta (1977) pointed out that stretching may not always be linear in the real world. Hence, here we explore nonlinear stretching.

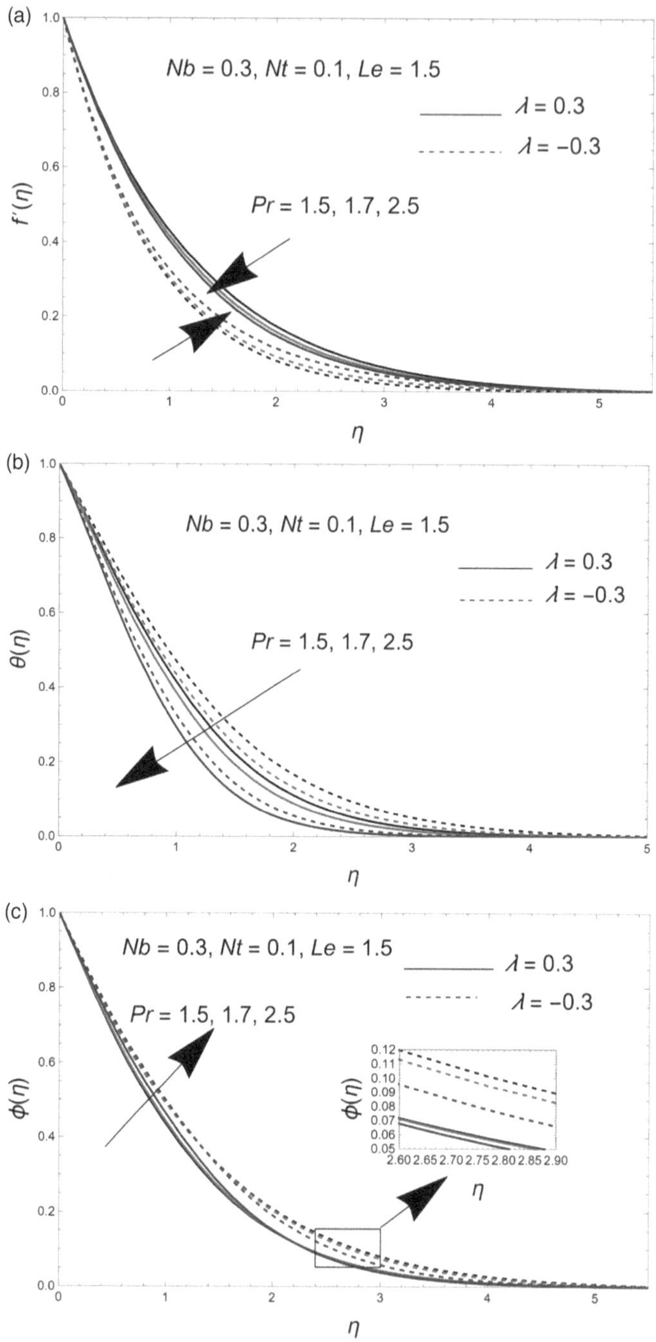

FIGURE 3.3 (a) Velocity, (b) temperature, and (c) concentration profiles for several values of the Prandtl number for buoyancy-aided/opposed flow.

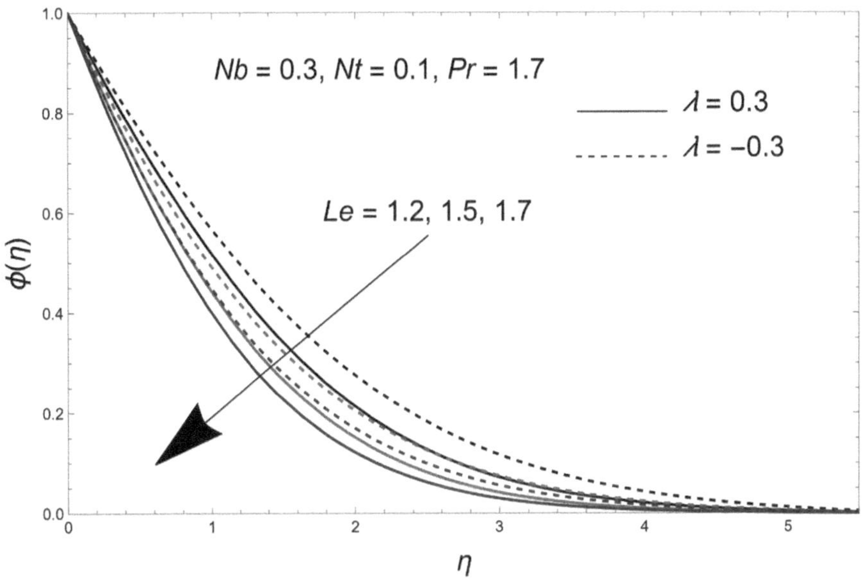

FIGURE 3.4 Concentration profiles for several values of Lewis number for buoyancy-aided/opposed flow.

Recently, Bisht and Maheshwari (2023) inspected magnetized nanofluid flow over a nonlinearly stretching surface. Santosh and Parida (2023) reported on the combined effects of a magnetic field and a heat source/sink on a nonlinearly stretching surface. Reddy and Mangamma (2023) discussed the mutual effects of thermal radiation and chemical reaction on nanofluid flow over a nonlinearly stretching sheet in the presence of a magnetic field and a nonuniform heat source. Triveni et al. (2023) examined the influences of magnetic fields and nonuniform heat sources on a Casson nanofluid flow over a nonlinearly stretching surface. Using a ternary hybrid non-Newtonian nanofluid, Naidu et al. (2024) reported on heat and mass transport for nanofluid flow due to a nonlinear stretching sheet in the presence of a magnetic field and Stefan blowing. Considering magneto-diffusion, Sewvandi et al. (2024) analyzed the nanofluid flow past a nonlinearly stretching sheet. These recent publications are representative of a larger body of research in this field.

Due to thermophoresis, nanoparticle flux becomes zero at the boundary. Considering this fact, we are interested in exploring nanofluid flow over a nonlinearly stretching surface having zero nanoparticle flux at the boundary; zero nanoparticle flux is generally used for passive control of flow. A two-phase model will be used for nanofluid flow that includes Brownian motion and thermophoresis.

3.2.1 Formulation of the Problem

The physical situation is shown in Figure 3.7. There, a schematic of the velocity boundary layer is provided, along with the concentration of nanoparticles within

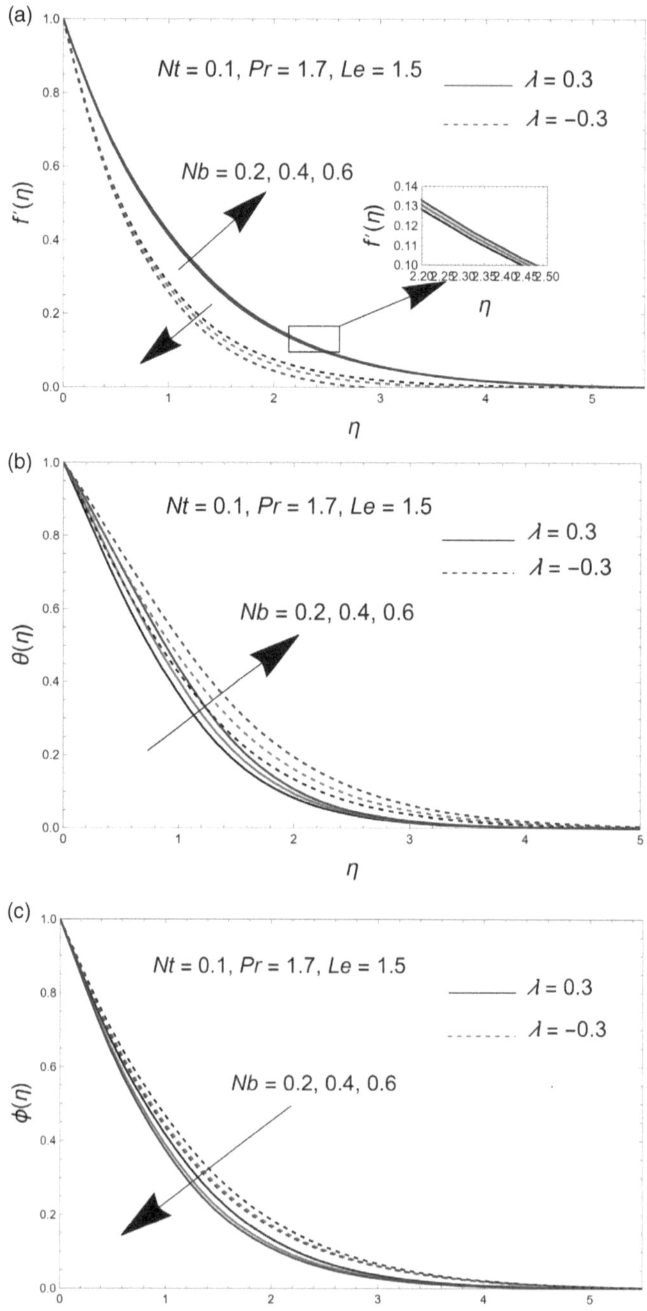

FIGURE 3.5 (a) Velocity, (b) temperature, and (c) concentration profiles for several values of the Brownian motion parameter for buoyancy-aided/opposed flow.

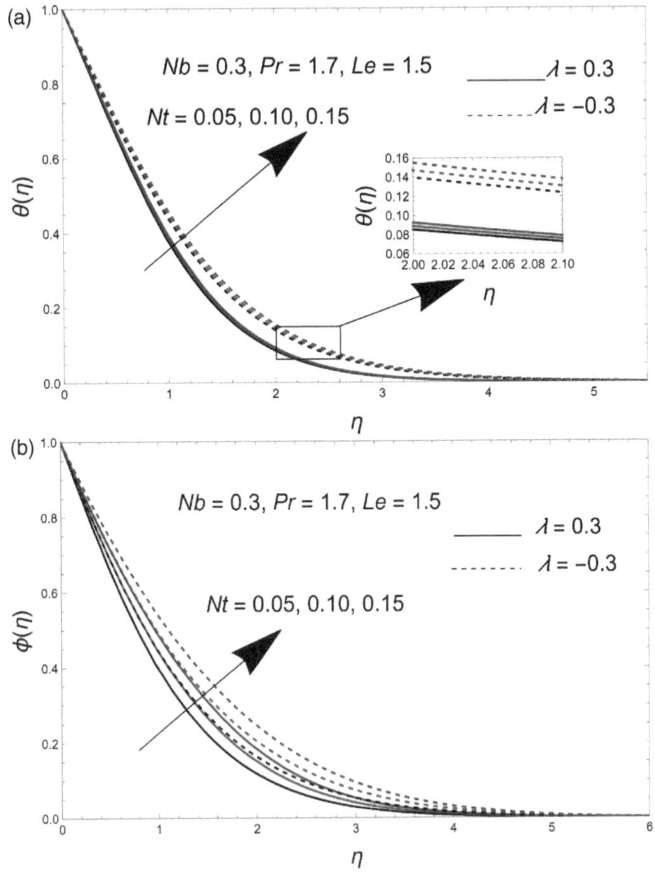

FIGURE 3.6 (a) Temperature and (b) concentration profiles for several values of thermophoresis parameter for buoyancy-aided/opposed flow.

TABLE 3.2
Values of [$f''(0)$, $\theta'(0)$, $\varphi'(0)$] with λ and with $Pr = 1.7$, $Le = 1.5$, $Nb = 0.3$, and $Nt = 0.1$

λ	$-\phi''(0)$	$-\theta'(0)$	$-\phi'(0)$
-0.3	1.165970	0.579683	0.603268
-0.1	1.053190	0.599122	0.631334
0.0	1.000480	0.607040	0.642513
0.1	0.949451	0.614208	0.652523
0.3	0.851268	0.626918	0.670047

TABLE 3.3
Values of $f''(0)$, $\theta'(0)$, and $\varphi'(0)$ with Pr and with $\lambda = 0.4$, $Le = 1.5$, $Nb = 0.3$ and $Nt = 0.1$

Pr	$-f''(0)$	$-\theta'(0)$	$-\varphi'(0)$
0.7	0.745968	0.430612	0.746649
1.7	0.803378	0.632828	0.677841
2.7	0.827968	0.744717	0.643293
3.5	0.839941	0.803137	0.626181

TABLE 3.4
Values of $f''(0)$, $\theta'(0)$, and $\varphi'(0)$ with Le and with $\lambda = 0.4$, $Pr = 1.7$, $Nb = 0.3$ and $Nt = 0.1$

Le	$-f''(0)$	$-\theta'(0)$	$-\varphi'(0)$
0.7	0.806772	0.693499	0.318736
1.2	0.804138	0.650069	0.560014
1.7	0.803052	0.623520	0.749023
2.5	0.802517	0.596891	0.994059

TABLE 3.5
Values of $f''(0)$, $\theta'(0)$, and $\varphi'(0)$ for Nb and $\lambda = 0.4$, $Pr = 1.7$, $Le = 1.5$ and $Nt = 0.1$

Nb	$-f''(0)$	$-\theta'(0)$	$-\varphi'(0)$
0.1	0.812423	0.749783	0.340447
0.5	0.794486	0.529285	0.744715
0.7	0.785830	0.438726	0.772993
1.2	0.765668	0.264253	0.801700

TABLE 3.6
Values of $f''(0)$, $\theta'(0)$, and $\varphi'(0)$for a variety of values of Nt with $\lambda = 0.4$, $Pr = 1.7$, $Le = 1.5$ and $Nb = 0.3$

Nt	$-f''(0)$	$-\theta'(0)$	$-\varphi'(0)$
0.1	0.803378	0.632828	0.677841
0.5	0.788464	0.518168	0.364870
0.9	0.774144	0.427829	0.228443
1.2	0.763878	0.372972	0.201052

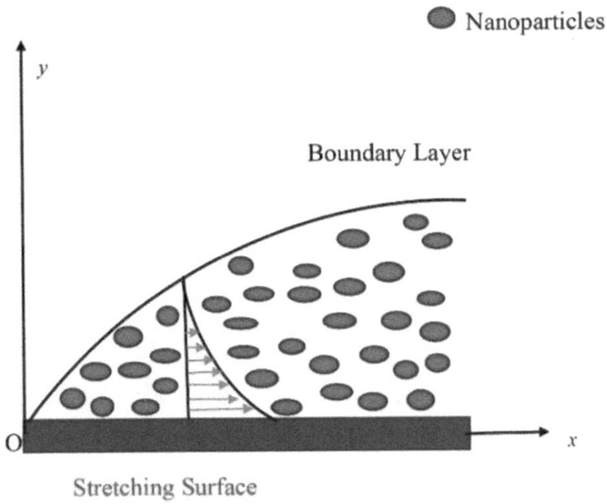

FIGURE 3.7 Sketch of the physical flow problem.

the boundary layer. To develop the governing equations, let (u,v) be the velocity components in the x and y directions, respectively.

Then, the continuity, momentum, energy, and species conservation equations governing the flow of nanofluid past a nonlinearly stretching sheet can be written as:

Mass conservation:

$$\frac{\partial u}{\partial x}+\frac{\partial \upsilon}{\partial y}=0. \tag{3.11}$$

Momentum conservation:

$$u\frac{\partial u}{\partial x}+\upsilon\frac{\partial u}{\partial y}=\nu\frac{\partial^2 u}{\partial y^2}. \tag{3.12}$$

Conservation of energy:

$$u\frac{\partial T}{\partial x}+\upsilon\frac{\partial T}{\partial y}=\frac{\kappa}{\rho c_p}\frac{\partial^2 T}{\partial y^2}+\tau\left[D_B\frac{\partial C}{\partial y}\frac{\partial T}{\partial y}+\frac{D_T}{T_\infty}\left(\frac{\partial T}{\partial y}\right)^2\right]. \tag{3.13}$$

Species conservation:

$$u\frac{\partial C}{\partial x}+\upsilon\frac{\partial C}{\partial y}=D_B\frac{\partial^2 C}{\partial y^2}+\frac{D_T}{T_\infty}\frac{\partial^2 T}{\partial y^2} \tag{3.14}$$

where $\nu = \mu/\rho$ is the kinematic viscosity, ρ is the fluid density, μ is the coefficient of fluid viscosity, the Brownian diffusion coefficient is D_B, the thermophoresis diffusion coefficient is D_T, T is the liquid temperature, and C is the concentration of the liquid, κ is the thermal conductivity, ρc_p is the heat capacitance of liquid, $\tau = (\rho c_p)_p/(\rho c_p)_p$, where $(\rho c_p)_p$ and $(\rho c_p)_f$ represent the specific heat capacities of the nanoparticle and base fluid respectively.

The appropriate boundary conditions for the problem are given by:

$$\text{at } y=0, u(x)=ax^n, \upsilon=0, T=T_w, D_B\frac{\partial C}{\partial y}+\frac{D_T}{T_\infty}\frac{\partial T}{\partial y}=0 \tag{3.15}$$

$$\text{as } y\to\infty,\ u(x)\to 0, T\to T_\infty, C\to C_\infty. \tag{3.16}$$

The surface temperature is T_w, and T_∞ is the uniform temperature of the free stream. C_w is the concentration of nanoparticles at the surface, and C_∞ is the nanofluid concentration away from the surface. To get self-similar solutions of Equations (3.11)–(3.14) subject to Equations (3.15)–(3.16), we introduce the following similarity transformations:

$$\eta = y\sqrt{\frac{a(n+1)}{2\nu}}\,x^{\frac{n-1}{2}}, u = ax^n f'(\eta), \upsilon = -\sqrt{\frac{a\nu(n+1)}{2}}\,x^{\frac{n-1}{2}}\left(f+\frac{n-1}{n+1}\eta f'\right) \tag{3.17}$$

$$\theta(\eta)=\frac{T-T_\infty}{T_w-T_\infty}, \phi(\eta)=\frac{C-C_\infty}{C_w-C_\infty}$$

where the similarity variable is η.

Upon substitution of (3.17) into Equations (3.12)–(3.14) and with the boundary conditions from Equations (3.15)–(3.16), we have the following, with dimensionless terms summarized in Table 3.7:

$$f'''+ff''-\left(\frac{2n}{n+1}\right)f'^2=0, \tag{3.18}$$

$$\frac{1}{Pr}\theta''+f\theta'+Nb\,\theta'\phi'+Nt\,\theta'^2=0, \tag{3.19}$$

TABLE 3.7
Values of $f''(0)$ for various values of nonlinear stretching parameter n

Nonlinear stretching parameter (n)	Hamad et al. (2011)	Ramya et al. (2018) with $\lambda = 0 = \delta$	Present result ($Le = 1$, $Pr = 6.8$)
0.0	0.6283	0.6283	0.6283
0.5	0.8901	0.8901	0.8901
1.0	1.0004	1.0005	1.0006
3.0	1.1489	1.1490	1.1503
10.0	1.2352	1.2352	1.2395

$$\phi'' + Le\, f\phi' + \frac{Nt}{Nb}\theta'' = 0, \tag{3.20}$$

$$f(0) = 0, f'(0) = 1, \theta(0) = 1, \phi'(0) + \frac{Nt}{Nb}\theta'(0) = 0, \tag{3.21}$$

$$f'(\eta) = 0, \theta(\eta) = 0, \phi(\eta) = 0 \text{ as } \eta \to \infty. \tag{3.22}$$

Multiple dimensionless numbers involved here are defined as follows:

$$Nb = (\rho c_p)_p D_B (C_w - C_\infty)/(\rho c_p)_f v,\ Nt = (\rho c_p)_p D_T (T_w - T_\infty)/(\rho c_p)_f T_\infty v,$$
$$Le = v/D_B,\ Pr = v\rho c_p/\kappa.$$

Here, Nb, Nt, Le, and Pr are the Brownian motion parameter, thermophoresis parameter, Lewis number, and Prandtl number, respectively.

3.2.2 Numerical Solutions and Discussion of the Results

Numerical computations have been carried out using the fourth order classical Runge-Kutta method with a shooting technique. For verifying the numerical code, results are compared with the published results for specific cases and are presented in Table 3.7. The table displays the values of ($f''(0)$) for the variations of nonlinear stretching parameter n for ordinary viscous fluids. Our results for viscous liquids under steady motion match with those of Hamad et al. (2011) and Ramya et al. (2018).

Fluid velocity is reduced with increasing values of the nonlinear stretching parameter n, as displayed in Figure 3.8a. But the temperature (see Figure 3.8b) and concentration (shown in Figure 3.9c) both rise with increasing values of nonlinear stretching parameter n. Higher variations in all cases are noted for lower values of n since as n $\to \infty$ $2n/(n+1)$ approaches 2 in Equation (3.18).

Fluid velocity also diminishes with rising values of the nonlinear stretching parameter n for values of n that are negative (Figure 3.9a). Both the temperature

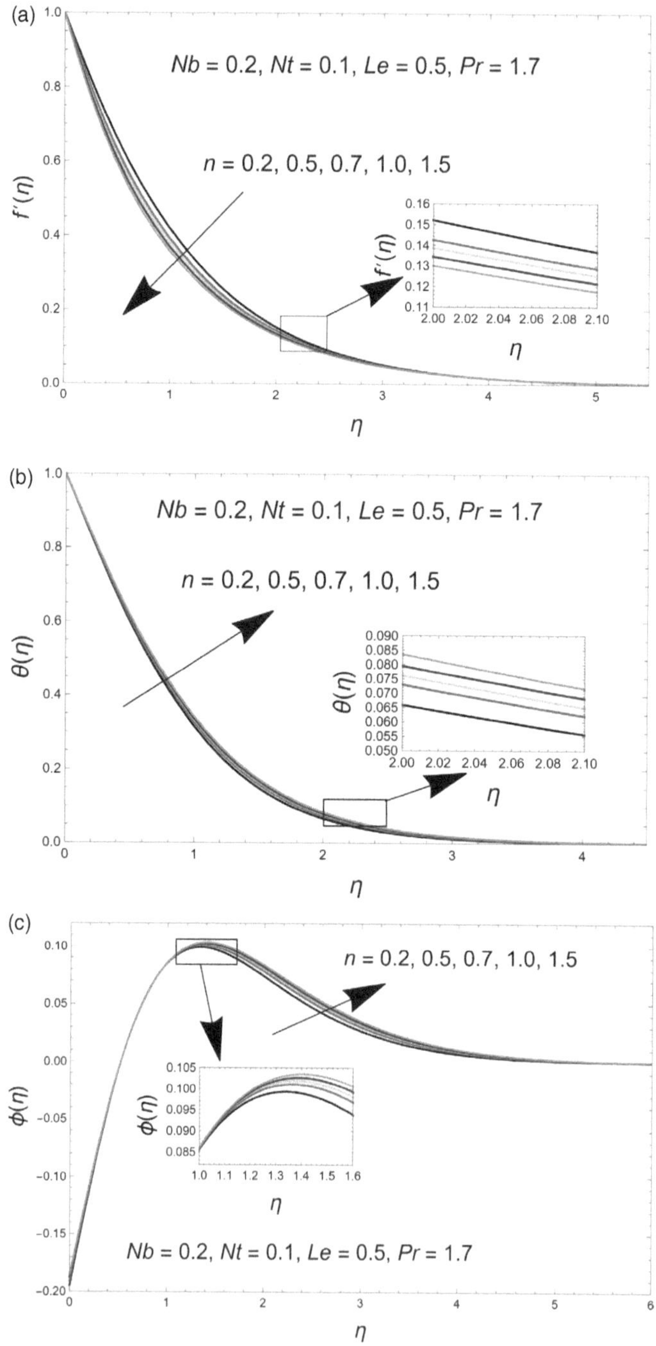

FIGURE 3.8 Variations of (a) velocity, (b) temperature, and (c) concentration for different values of non-linear stretching parameter n.

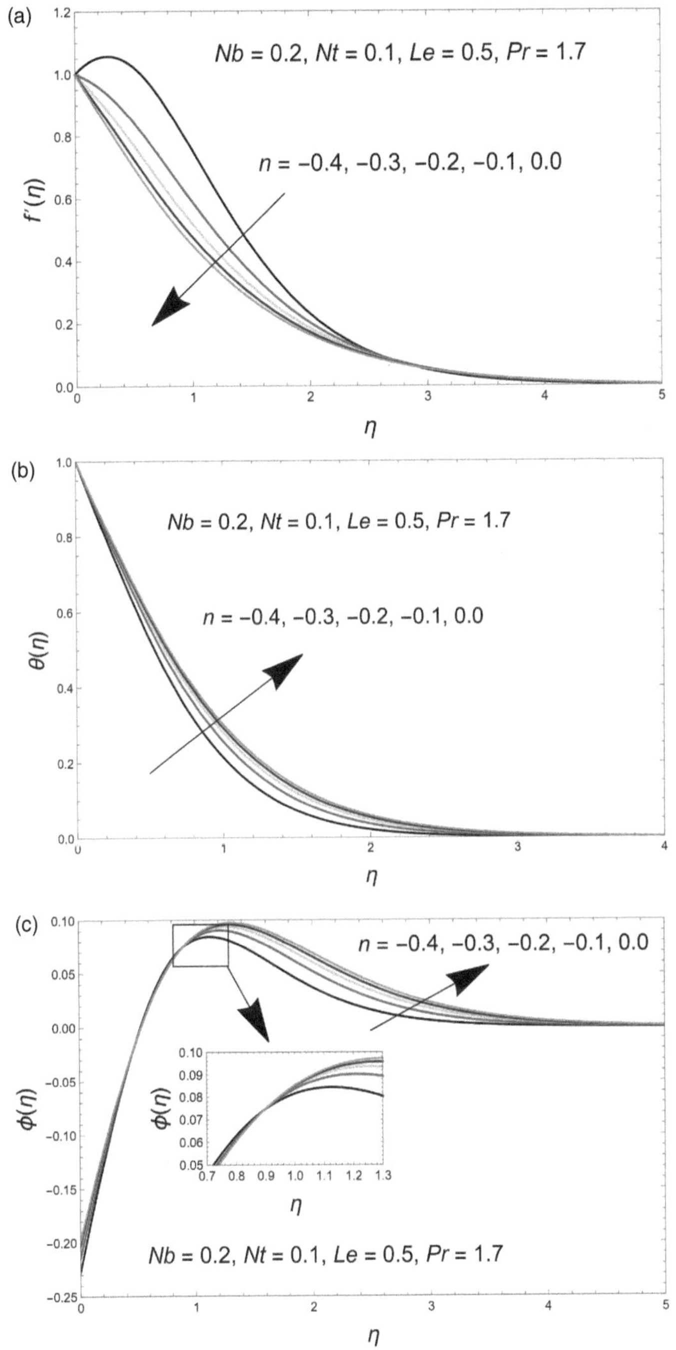

FIGURE 3.9 Variations of (a) velocity, (b) temperature, and (c) concentration for different (negative) values of non-linear stretching parameter n.

(Figure 3.9b) and concentration (Figure 3.9c) increase with rising values of *n* for values of *n* that are negative.

Though concentration initially rises near the surface, deeper into the boundary layer, it declines. Additionally, larger values of the Brownian parameter lead to lower values in concentration (Figure 3.10a). Variation of concentration with Lewis number is shown in Figure 3.10b. Near the wall and the edge of the boundary layer, concentrations are the same; however in the middle of the boundary layer, there is a clear impact of the Lewis number.

Temperature rises significantly for larger values of thermophoresis number, *Nt* (Figure 3.11a). Far from the surface, concentration also rises with the rise in thermophoresis number *Nt* (Figure 3.11b). However, near the surface, a diminishing effect of concentration is noted with the rise in thermophoresis parameter *Nt* (Figure 3.11b).

Temperature diminishes significantly with an increase in the Prandtl number (Figure 3.12a). The concentration has its largest values at intermediate locations within the boundary layer (Figure 3.12b). The thermal boundary layer thickness is reduced, but the solute boundary layer thickness rises with the rising values of *Pr*.

For increasing nonlinear stretching parameter n (>0), the surface shear stress diminishes, but the temperature gradient at the surface rises (Table 3.8). For all values of *n* greater than zero, the surface shear stress is negative, which is expected for flow past a stretching sheet. The temperature gradient at the surface is also negative, which indicates that heat is transferred from the surface to the fluid. On the other hand, the concentration gradient at the surface is positive for all values of n greater than zero, which indicates that mass is absorbed at the surface. The concentration gradient at the surface decreases with rising values of the stretching parameter n (>0).

With rising values of Brownian motion parameter *Nb*, the concentration gradient at the surface decreases significantly, as seen in Table 3.9.

The temperature gradient rises with an increase in the thermophoresis parameter *Nt* (Table 3.10). The concentration gradient at the surface significantly rises with the thermophoresis parameter *Nt* (Table 3.10). Because of thermophoresis, nanoparticles are shifted from hot zones to cold zones.

3.3 NANOFLUID FLOW PAST AN EXPONENTIAL STRETCHING SHEET

Currently, nanofluid flow, heat, and mass transfer past a stretching sheet have received considerable attention for their widespread applications and relevance in several technological processes. Most of the previously published investigations deal with the study of nanofluid flow past a stretching surface where the velocity of the stretching sheet varies linearly with distance from the fixed origin. In the real world, stretching may not be linear, and the stretching velocity may vary exponentially. With this in mind, many researchers have studied nanofluid flow due to an exponentially stretching surface. Recently, with thermophoretic particle deposition,

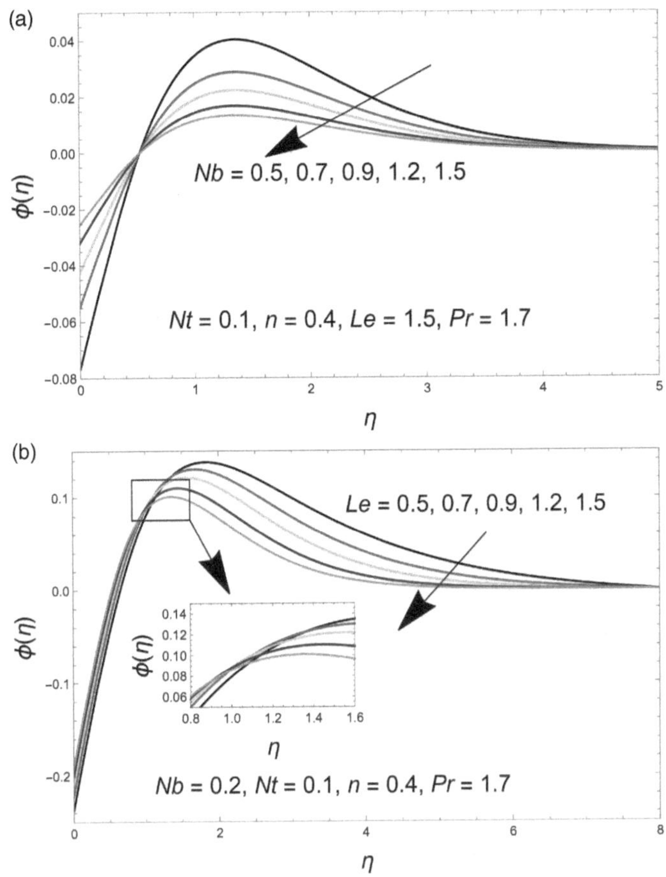

FIGURE 3.10 Variations of concentration for different values of (a) Brownian motion parameter *Nb* and (b) Lewis numbers *Le*.

Prasannakumara et al. (2023) examined bioconvective nanofluid flow due to an exponentially stretching sheet. Considering viscous dissipation and Joule heating, Nandi and Kumbhakar (2023) studied hybrid nanofluid flow over an exponentially stretching sheet and reported the impact of an obliquely applied magnetic field. Kathyayani and Prakash Gowd (2023) discussed the combined effects of various shape factors, activation energy, and a heat source in hybrid nanofluid flow over an exponentially stretching sheet. Mahmood and Khan (2023) explored the impacts of nanoparticle aggregation on nanofluid flow past an exponentially stretching sheet embedded in a porous medium. Khader et al. (2023) analyzed the effects of viscous dissipation on non-Newtonian nanofluid flow over an exponentially stretching surface. They also analyzed the effects of a Darcy-Forchheimer porous medium.

Considering activation energy and motile microorganisms, Sankari et al. (in press) investigated bio-convective Casson nanofluid flow over an exponentially stretching

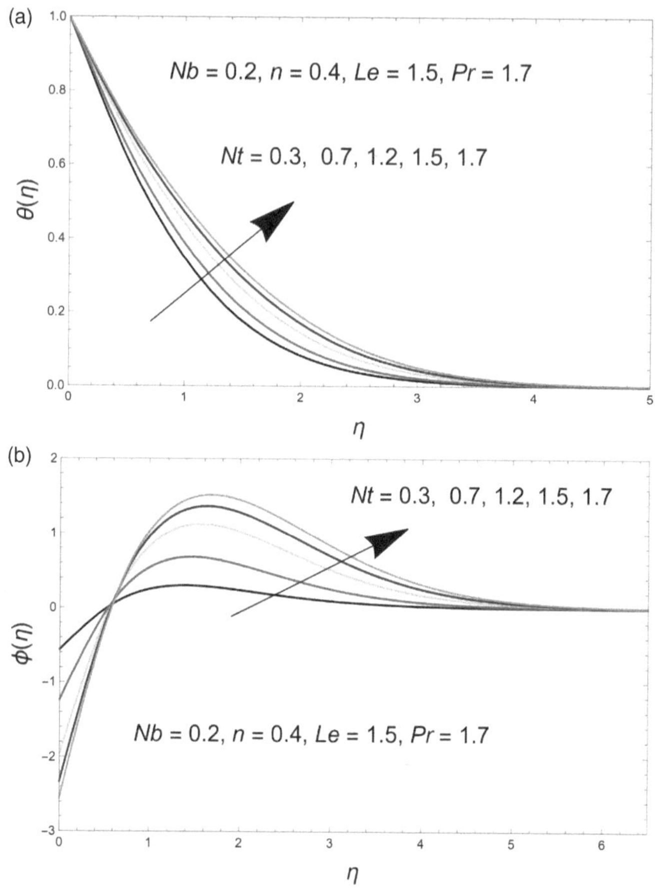

FIGURE 3.11 Variations of (a) temperature and (b) concentration for different values of thermophoresis parameter *Nt*.

sheet. They obtained analytic solutions using the Homotopy Analysis Method. Considering chemical reactions and heat generation/absorption, Preetham and Kumbinarasaiah (2024) examined nanofluid flow due to an exponentially stretching surface in the presence of a magnetic field. They used a two-phase model in their study and applied the Haar wavelet method to obtain a numerical solution. Mandal and Pal (2024) obtained dual solutions for nanofluid flow due to an exponentially stretching/shrinking surface embedded in a Darcy-Forchheimer porous medium in the presence of thermal radiation and a magnetic field. They presented a stability analysis. Naveed Khan et al. (2024) studied second-grade nanofluid flow past an exponentially stretching sheet near a stagnation point in the presence of chemical reactions. Razzaq and Farooq (2024) obtained non-similar solutions for boundary layer hybrid nanofluid flow over an exponentially stretching/shrinking surface in the presence of a magnetic field, with thermal radiation and viscous dissipation.

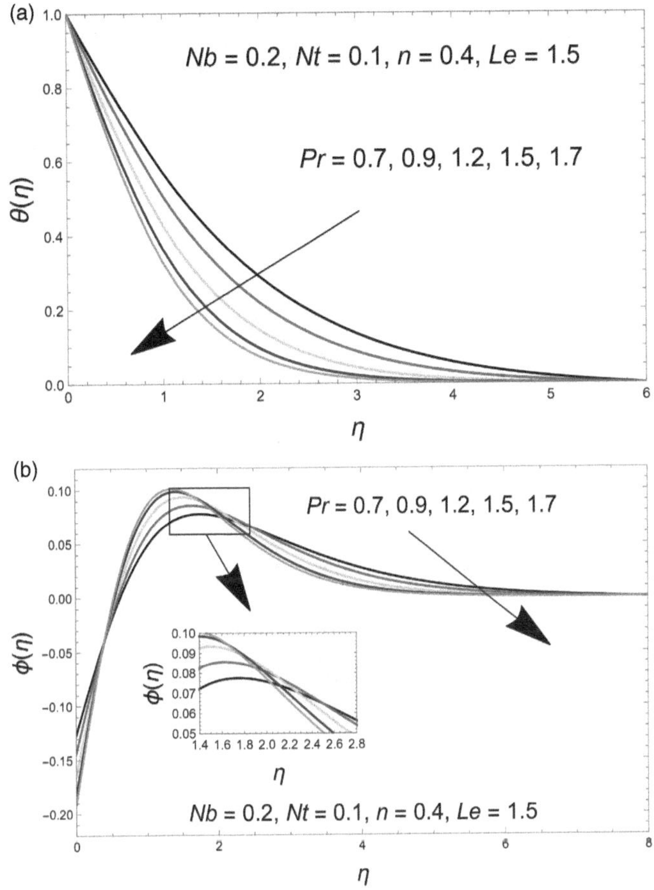

FIGURE 3.12 Variations of (a) temperature and (b) concentration for different values of Prandtl number, *Pr*.

TABLE 3.8
Values of $f''(0)$, $\theta'(0)$ and $\varphi'(0)$ for various values of nonlinear stretching parameter n with $Nb = 0.2$, $Nt = 0.1$, and $Le = 0.5$, $Pr = 1.7$

n	$f''(0)$	$\theta'(0)$	$\varphi'(0)$
0.1	-0.706624	-0.857710	0.428855
0.4	-0.856472	-0.836302	0.418151
1.0	-1.000480	-0.816142	0.408071
1.6	-1.071250	-0.806412	0.403206

TABLE 3.9
Values of $\varphi'(0)$ for various values of Brownian motion parameter Nb with $Nt = 0.1$, $n = 0.4$, $Le = 0.5$ and $Pr = 1.7$

Nb	$\varphi'(0)$
0.2	0.418150
0.6	0.139380
1.0	0.083630
2.0	0.041815
2.6	0.032166
3.0	0.027877

TABLE 3.10
Values of $\varphi'(0)$ for various values of thermophoresis parameter Nt with $Nb = 0.1$, $n = 0.4$, $Le = 0.5$, and $Pr = 1.7$

Thermophoresis Parameter, Nt	$\theta'(0)$	$\varphi'(0)$
0.1	-0.836302	0.148151
0.4	-0.815389	1.630780
0.8	-0.787241	3.148960
1.0	-0.773067	3.865330
1.6	-0.730241	5.841930
2.2	-0.687180	7.558980
2.6	-0.658517	8.560720
3.0	-0.630040	9.450600

The previously mentioned research works considered a static free stream. Some situations occur in reality where the free stream moves and is unsteady, as discussed in Magyari and Keller (1999). Here, we are interested in presenting the results obtained for nanofluid flow past an exponentiall y stretching surface in the presence of a moving free stream. It is expected that these results will provide valuable information and will serve as complements to the previous studies.

3.3.1 Mathematical Formulation

Consider the flow of an incompressible viscous nanofluid past a flat sheet coinciding with the plane $y = 0$. The flow is confined to $y > 0$. Two equal and opposite forces are applied along the x-axis so that the wall is stretched, keeping the origin fixed. The continuity, momentum, energy, and concentration equations can be written as:

Conservation of mass:

$$\frac{\partial u}{\partial x}+\frac{\partial u}{\partial y}=0. \tag{3.23}$$

Conservation of momentum:

$$u\frac{\partial u}{\partial x}+\upsilon\frac{\partial u}{\partial y}=u_{\infty}\frac{du_{\infty}}{dx}+\nu\frac{\partial^2 u}{\partial y^2}. \tag{3.24}$$

Conservation of energy:

$$u\frac{\partial T}{\partial x}+\upsilon\frac{\partial T}{\partial y}=\frac{\kappa}{\rho c_p}\frac{\partial^2 T}{\partial y^2}+\tau\left[D_B\frac{\partial C}{\partial y}\frac{\partial T}{\partial y}+\frac{D_T}{T_{\infty}}\left(\frac{\partial T}{\partial y}\right)^2\right]. \tag{3.25}$$

Species conservation:

$$u\frac{\partial C}{\partial x}+\upsilon\frac{\partial C}{\partial y}=D_B\frac{\partial^2 C}{\partial y^2}+\frac{D_T}{T_{\infty}}\frac{\partial^2 T}{\partial y^2}. \tag{3.26}$$

Here, u and v are the components of velocity respectively in the x and y directions, $\nu = \mu/\rho$ is the kinematic viscosity, ρ is the fluid density, μ is the fluid viscosity, κ is the thermal conductivity of the fluid, the Brownian diffusion coefficient is D_B, the thermophoresis diffusion coefficient is D_T, T is the liquid temperature, T_{∞} stands for constant temperature of the nanofluid in free stream, C is the concentration of the liquid, C_{∞} stands for constant concentration of the nanofluid in the free stream, ρc_p is the heat capacitance of liquid. $\tau = (\rho c_p)_p/(\rho c_p)_f$; $(\rho c_p)_p$ and $(\rho c_p)_f$ represent specific heat capacities of the nanoparticles and base fluid correspondingly.

The appropriate boundary conditions for the problem are given by

$$u=U_w \quad \upsilon=0 \; T=T_w \quad C=C_w \quad \text{at } y=0 \tag{3.27}$$

$$u\rightarrow u_{\infty}(x), T\rightarrow 0,\, C\rightarrow 0 \text{ as } y\rightarrow\infty. \tag{3.28}$$

Here, $U_w = a\exp(x/L)$ is the stretching velocity where $a > 0$ is the stretching constant, T_w is the temperature at the sheet, C_w is the nanoparticle volume fraction at the wall. The term $u_{\infty} = b\exp(x/L)$ is the free stream velocity where $b>0$ is a constant, and L is a constant having dimension of length.

To get self-similar solutions of Equations (3.24)–(3.26) subject to Equations (3.27)–(3.28), we introduce the following similarity transformations:

$$\psi = \sqrt{2\nu La}\, f(\eta)\exp\left(\frac{x}{2L}\right),\quad \eta = y\sqrt{\frac{a}{2\nu L}}\exp\left(\frac{x}{2L}\right), \tag{3.29}$$

$$\theta(\eta) = \frac{T - T_\infty}{T_w - T_\infty},\quad \phi(\eta) = \frac{C - C_\infty}{C_w - C_\infty}$$

where ψ is the stream function given by $u = \partial\psi/\partial y$, $v = -\partial\psi/\partial x$.

It should be noted that Equation (3.23) is automatically satisfied by the above transformations, and upon substitution of Equation (3.29) into Equations (3.24)–(3.26), the governing equations reduce to

$$f''' + ff'' - 2f'^2 + 2\lambda^2 = 0, \tag{3.30}$$

$$\theta'' + Pr(f\theta' + Nb\,\theta'\phi' + Nt\,\theta'^2) = 0, \tag{3.31}$$

$$\phi'' + Le\,f\phi' + \frac{Nt}{Nb}\theta'' = 0. \tag{3.32}$$

Multiple dimensionless numbers involved here are defined as follows:

$$\lambda = b/a,\ Nb = \tau D_B (C_w - C_\infty)/\nu,\ Nt = \tau D_T (T_w - T_\infty)/T_\infty \nu,\ Le = \nu/D_B,\ Pr = \mu c_p/\kappa.$$

Here, λ is the velocity ratio parameter. The boundary conditions take the following form:

$$f' = 1, f = 0, \theta = 1, \phi = 1 \text{ at } \eta = 0 \tag{3.33}$$

and

$$f' \to \lambda, \theta \to 0, \phi \to 0 \text{ as } \eta \to \infty \tag{3.34}$$

where the prime denotes differentiation with respect to η.

3.3.2 Numerical Solutions and Analysis of the Results

The numerical calculations to a desired level of accuracy were carried out using the shooting method for different values of dimensionless parameters. To judge the accuracy of the numerical scheme, the present results for an ordinary viscous Newtonian fluid are compared to the results of Magyari and Keller (1999), Elbashbeshy (2001), Sahoo and Poncet (2011), and Ghosh and Mukhopadhyay (2017). The comparison

TABLE 3.11
Values of $f''(0)$ for ordinary viscous Newtonian fluids

Magyari and Keller (1999)	Elbashbeshy (2001)	Sahoo and Poncet (2011)	Ghosh and Mukhopadhyay (2017)	Present study
-1.28180	-1.28181	-1.28181	-1.28181	-1.28181

shows excellent agreement with their results (Table 3.11). Other outcomes of the numerical simulation are presented in Figures 3.13–3.17.

In Figure 3.13a, velocity profiles of the nanofluid are shown for different values of the velocity ratio parameter λ. The velocity curves show that the rate of transport decreases with the increasing distance (η). With increasing λ, the velocity of the nanofluid increases. With increasing λ, the temperature decreases, as shown in Figure 3.13b. Away from the surface, the concentration decreases for an increase in the velocity ratio parameter λ (Figure 3.13c).

Figures 3.14a–b display the variations of the nanofluid's nondimensional temperature and concentration for variations of the Brownian motion parameter, *Nb*. Figure 3.14a indicates that the temperature rises with increasing values of the Brownian motion parameter *Nb*. But the concentration decreases with the rise in *Nb* (Figure 3.14b). Concentration overshoot is observed near the surface for Brownian motion of nanoparticles. The thermal boundary layer thickness rises but the solute boundary layer thickness reduces.

Figures 3.15a–b show the variations of temperature and concentration of the nano-liquid for variations of the thermophoresis parameter, *Nt*. Figures 3.15a–b demonstrate that the temperature (Figure 3.15a) and the concentration (Figure 3.15b) of the nano-liquid both increase for increasing values of the thermophoresis parameter, *Nt*. Physically, the thermophoresis force increases with rising values of *Nt*, causing the nanoparticles to move from hot to cold areas. For this reason, the temperature of the nanofluid and the concentration both are found to increase.

Figures 3.16a–b describe the variations of nondimensional temperature and concentration of the fluid for variations of the Prandtl number. Figure 3.16a indicates that the temperature of the fluid decreases for growing values of *Pr*. Actually, the thickness of the thermal boundary layer increases due to the higher thermal conductivity of nanofluids and the consequent lower Prandtl number. Figure 3.16b shows that the concentration of the nanofluid initially grows for increasing values of the Prandtl number. Still, away from the surface, the concentration diminishes for higher values of *Pr* (Figure 3.16b).

Figure 3.17 describes the concentration of the nano-liquid for various Lewis numbers. Figure 3.17 shows that the concentration of the nano-liquid decreases for rising values of *Le*. Actually, the concentration boundary layer thickness decreases, and the mass transfer increases as the Lewis number rises. When the Lewis number increases, the Brownian diffusion coefficient D_B decreases. As a result, the thickness of the concentration boundary layer decreases.

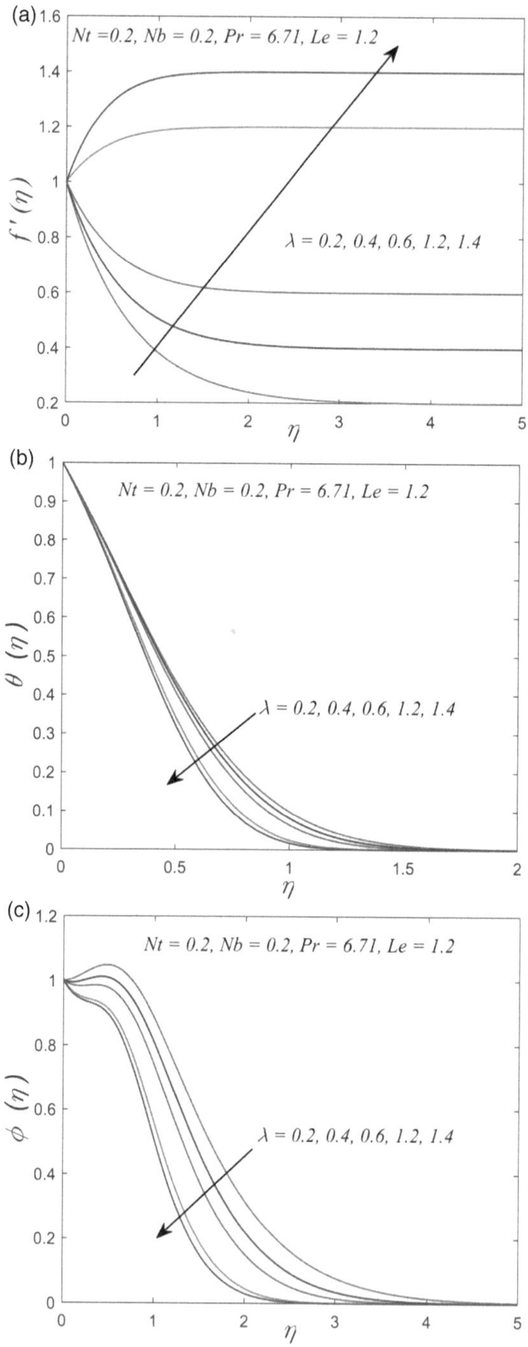

FIGURE 3.13 Effects of velocity ratio parameter λ on (a) velocity, (b) temperature, and (c) concentration profiles.

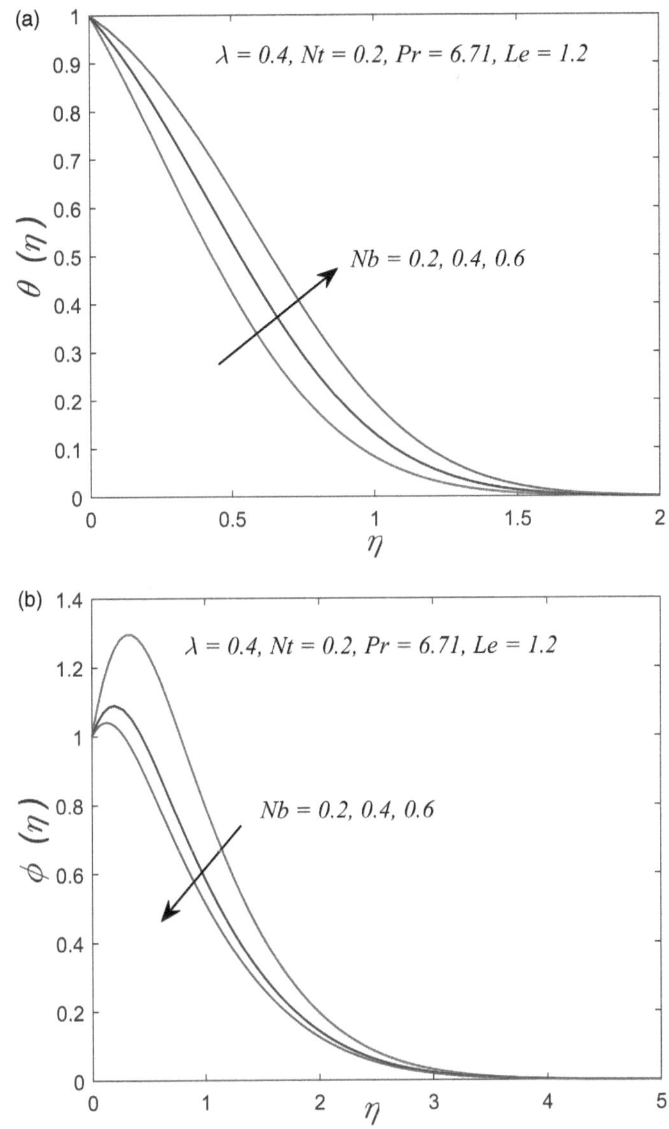

FIGURE 3.14 Effects of Brownian motion parameter Nb on (a) temperature and (b) concentration profiles.

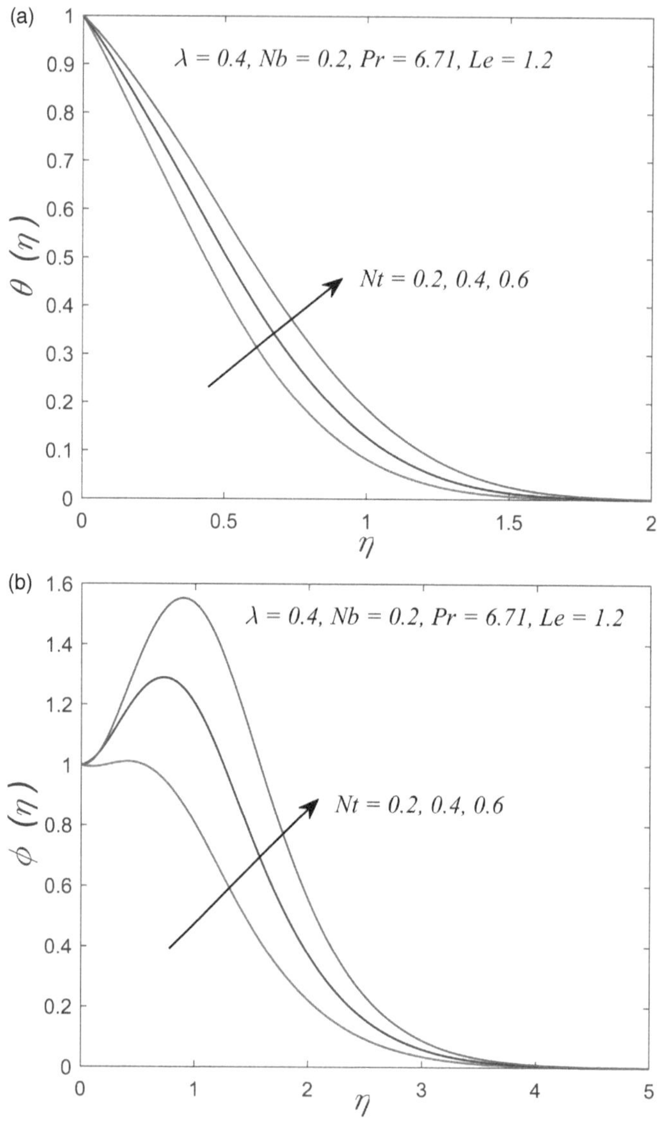

FIGURE 3.15 Effects of thermophoresis parameter Nt on (a) temperature and (b) concentration profiles.

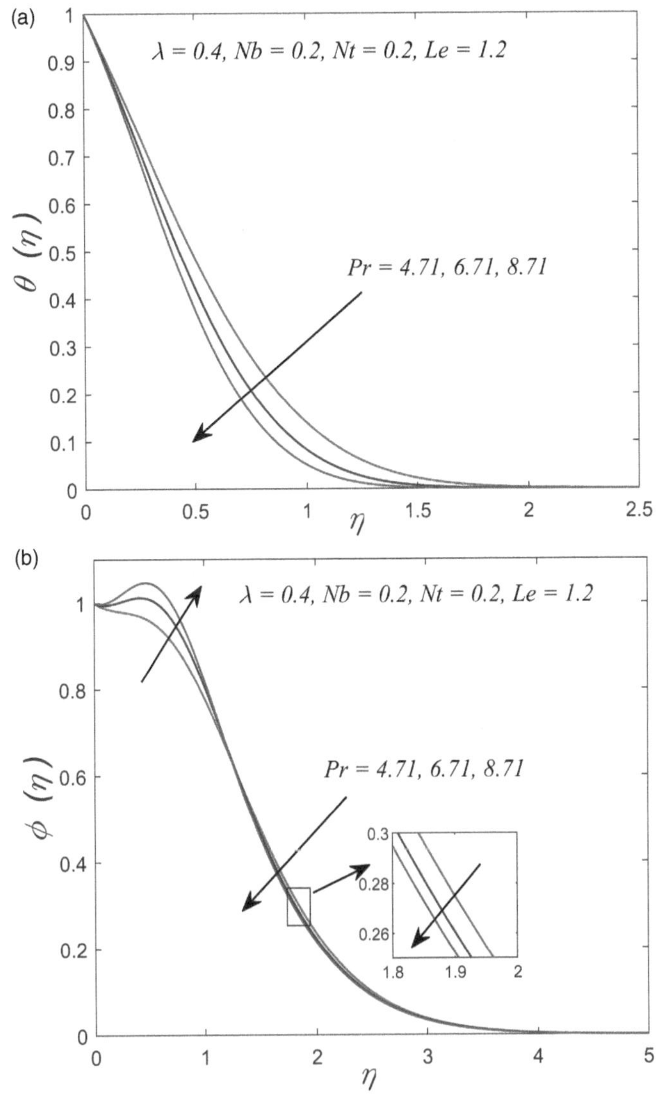

FIGURE 3.16 Effects of Prandtl number Pr on (a) temperature and (b) concentration profiles.

3.4 NANOFLUID FLOW PAST AN UNSTEADY STRETCHING SHEET

Generally, the flow, temperature, and concentration fields are considered to be steady for studies related to nanofluid flow past a stretching surface. However, some situations arise where the situation may be unsteady due to impulsive stretching of the flat surface or by a step change of the surface temperature or wall concentration, for example.

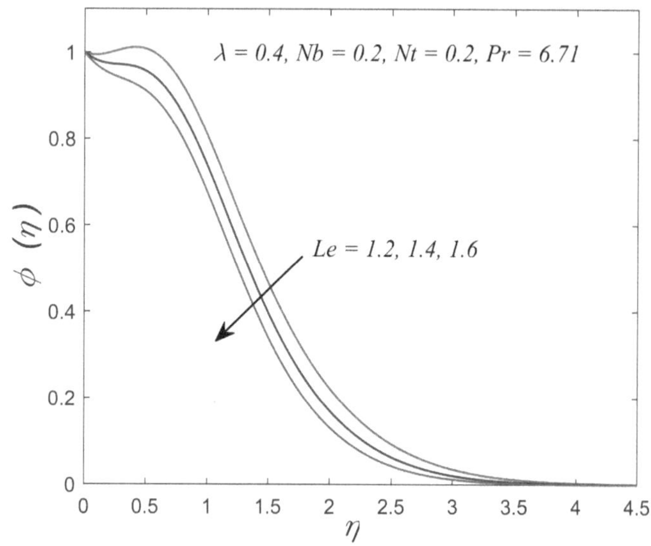

FIGURE 3.17 Effects of Lewis number *Le* on concentration profiles.

Mandal and Shit (2021) presented entropy generation for three-dimensional unsteady flow of a Williamson nanofluid due to a convectively heated stretching sheet. They reported the effects of the magnetic field on flow and heat transport. Makkar et al. (2021) examined Casson nanofluid flow over an unsteady stretching surface in the presence of a magnetic field with the help of Buongiorno's model. They discussed the effects of chemical reactions on flow fields. Ramanjini et al. (2022) analyzed mixed convective Williamson nanofluid flow past an unsteady radially stretching Riga plate in the presence of thermal radiation. Mahato et al. (2022) investigated the stagnation point flow of a Casson nanofluid past an unsteady stretching surface in a porous medium in the presence of an inclined magnetic field. They discussed the effects of homogeneous/heterogeneous chemical reactions. They also calculated the generated entropy. Recently, considering water-based nanofluids, Naseem and Shahzad (2023) investigated thin film flow past an unsteady stretching sheet. They also calculated the generated entropy. Hayat and Shahzad (2023) inspected nanofluid flow containing Au nanoparticles due to an unsteady radially stretching surface. Konai et al. (2023) highlighted the effects of Stefan blowing on a Casson nanofluid flow past an unsteady stretching surface. They also reported the effects of Brownian motion and thermophoresis.

The problems that have been considered in this chapter are based on a two-phase model for nanofluid that accounts for the effects of Brownian motion and thermophoresis. However, the thermophysical properties of nanoparticles are disregarded in the model. Moreover, this model is applicable only for 10nm–100 nm-sized nanoparticles. But for nanoparticles smaller than 10nm, one has to apply a single-phase model for the nanofluid which incorporates the effects of nanoparticles on thermophysical fluid properties. Thermophysical fluid properties are very important to flow, heat transfer, and concentration. With this in mind, an attempt has been made in this section to explore Cu-water nanofluid flow, heat transfer, and mass transfer past an unsteady stretching surface in the presence of thermal radiation.

3.4.1 Mathematical Formulation of the Problem

We consider laminar boundary-layer flow, heat, and mass transfer of incompressible Cu-water nanofluid over an unsteady stretching sheet. We assume that for time $t < 0$, the fluid, heat, and mass flows are steady. The unsteady fluid, heat transfer, and mass transfer start at $t = 0$. At the origin ($x = 0$, $y = 0$), the sheet emerges out of a slit and moves with non-uniform velocity $u(x,t) = bx/(1\text{-}at)$ where b and a are positive constants with dimensions $(\text{time})^{-1}$ $b$is the initial stretching rate and $b/(1\text{-}at)$ is the effective stretching rate which increases with time. Generally, in polymer extrusion problems, the properties of the material of the extruded sheet vary with time.

The governing equations of such type of flow are, in the usual notation,

$$\frac{\partial u}{\partial x}+\frac{\partial \upsilon}{\partial y}=0, \tag{3.35}$$

$$\frac{\partial u}{\partial t}+u\frac{\partial u}{\partial x}+\upsilon\frac{\partial u}{\partial y}=\frac{\mu_{nf}}{\rho_{nf}}\frac{\partial^2 u}{\partial y^2}. \tag{3.36}$$

$$\frac{\partial T}{\partial t}+u\frac{\partial T}{\partial x}+\upsilon\frac{\partial T}{\partial y}=\frac{k_{nf}}{\left(\rho c_p\right)_{nf}}\frac{\partial^2 T}{\partial y^2}-\frac{1}{\left(\rho c_p\right)_{nf}}\frac{\partial q_r}{\partial y}, \tag{3.37}$$

$$\frac{\partial C}{\partial t}+u\frac{\partial C}{\partial x}+\upsilon\frac{\partial C}{\partial y}=D\frac{\partial^2 C}{\partial y^2}. \tag{3.38}$$

Here, the viscous dissipation term in the energy equation is neglected (as the fluid velocity is low), and u and υ are the components of velocity respectively in the x and y directions, ρ_{nf}, μ_{nf} denote the density and viscosity of nanofluid, $(\rho c_p)_{nf}$ is the effective heat capacitance, and k_{nf} is the thermal conductivity of the nanofluid.

In these equations, T is the temperature, C is the concentration, D is the diffusion coefficient, and $q_r=-\frac{4\sigma^*}{3k^*}\frac{\partial T^4}{\partial y}$ is the radiative heat flux, σ^* denotes Stefan–Boltzmann constant, and k^* represents the mean absorption coefficient of the nanofluid. Further, expanding T^4 about T_∞ in a Taylor series, one can get $T^4 \cong 4T_\infty^{\,3}\,T\text{-}3\,T_\infty^{\,4}$.

In this single-phase fluid model, Cu nanoparticles are deposited in the ordinary base fluid water to prepare the nanofluid; the thermo-physical properties of Cu nanoparticles and water are given by Konai and Mukhopadhyay (2024).

$$\mu_{nf} = \frac{\mu_f}{(1-\phi)^{2.5}}, \tag{3.39a}$$

$$\rho_{nf} = (1-\phi)\rho_f + \phi\rho_s, \tag{3.39b}$$

$$\frac{k_{nf}}{k_f} = \frac{(k_s + 2k_f) - 2\phi(k_f - k_s)}{k_s + 2k_f + \phi(k_f - k_s)}, \tag{3.39c}$$

$$(\rho c_p)_{nf} = (1-\phi)(\rho c_p)_f + \phi(\rho c_p)_s. \tag{3.39d}$$

Here, φ is the volume fraction of the solid nanoparticles, k_f and k_s are, respectively, the thermal conductivities of base fluid and nanoparticles. The subscripts *nf*, *f*, and *sp* stand for nanofluid, base fluid, and solid nanoparticles, respectively.

For small-sized nanoparticles, D depends on the substance of the particles but this dependence becomes negligibly small for nanoparticles of diameter 7–10 nm. Here, we have considered Cu nanoparticles having a diameter of 7–10nm, so D has no dependence on solid volume fraction φ (Konai and Mukhopadhyay, 2024).

The appropriate boundary conditions for the problem are given by

$$u = u(x,t), \upsilon = 0, T = T_w, C = C_w \text{ at } y = 0, \tag{3.40a}$$

$$u \to 0, T \to T_\infty, C \to C_\infty \quad \text{as } y \to \infty. \tag{3.40b}$$

The temperatures at the surface and free stream are T_w and T_∞ (constant), respectively; the concentration of the nanofluid at the surface and in the free stream are C_w and C_∞ respectively.

We now introduce the following relations for u, υ, and θ, H as

$$u = \frac{\partial \psi}{\partial y}, v = \frac{\partial \psi}{\partial y}, \theta = \frac{T - T_\infty}{T_w - T_\infty}, \; H(\eta) = \frac{C - C_w}{C_w - C_\infty} \tag{3.41a}$$

where ψ is the stream function.

Let us introduce,

$$\psi = x\sqrt{\frac{\nu_{nf} b}{1-\alpha t}} f(\eta)\,,\ \eta = y\sqrt{\frac{b}{\nu_{nf}(1-\alpha t)}}. \tag{3.41b}$$

With the help of the above relations, the governing equations reduce to

$$\frac{1}{(1-\phi)^{2.5}\left(1-\phi+\phi\frac{\rho_s}{\rho_f}\right)} f''' + ff'' - f'^2 - A\left(f' + \frac{\eta}{2} f''\right) = 0, \tag{3.42}$$

$$\frac{1}{Pr}\frac{1}{\left(1-\phi+\phi\frac{(\rho c_p)_s}{(\rho c_p)_f}\right)}\left(\frac{k_{nf}}{k_f} + \frac{4}{3}R\right)\theta'' + f\theta' - \frac{A}{2}\eta\theta' = 0, \tag{3.43}$$

$$\frac{1}{Le} H'' + f H' - \frac{A}{2}\eta H' = 0. \tag{3.44}$$

Multiple dimensionless numbers involved here are defined as follows:

$$A = \alpha/b,\ R = 4T_\infty^{\ 3}\sigma^*/k^*k_f,\ Le = \nu_f/D,\ Pr = \nu_f(\rho c_p)_f/k_f$$

Here, A, R, Le, and Pr are the unsteadiness parameter, radiation parameter, Lewis number, and Prandtl number, respectively.

$$f'(0) = 1\,,\ f(0) = 0\,,\ \theta(0) = 1\,,\ H(0) = 1 \ \text{ when } \ \eta = 0. \tag{3.45}$$

$$f'(\eta) = 0\,,\ \theta(\eta) = 0\,,\ H(\eta) = 0 \text{ as } \eta \to \infty. \tag{3.46}$$

3.4.2 Numerical Solutions and Discussion of the Results

Using the shooting technique with the Runge-Kutta method, the above equations (3.42) to (3.44) subject to the boundary conditions (3.45)–(3.46) are numerically solved in order to validate the method comparison of present study with an ordinary viscous incompressible fluid ($\varphi = 0$) with available results of Sharidan et al. (2006) and Chamkha et al. (2010) in Table 3.12 show that the results agree well.

Another comparison corresponding to the data of temperature gradient at the surface for a steady flow of Cu-water nanofluid in the absence of thermal radiation is made with Wang (1989) and Rashidi et al. (2014) and presented in Table 3.13 which also shows the accuracy of our numerical data.

TABLE 3.12
The values of $f''(0)$ for various values of unsteadiness parameter A ordinary viscous incompressible fluid ($\varphi = 0$)

A	Sharidan et al. (2006)	Chamkha et al. (2010)	Present study
0.8	-1.261042	-1.261512	-1.261479
1.2	-1.377722	-1.378052	-1.261479

TABLE 3.13
Comparison of $-\theta'(0)$ for various values of Prandtl number Pr for Cu-water nanofluid

Pr	Wang (1989)	Rashidi et al. (2014) (N, λ and $\varphi = 0$)		Present results in absence A, R, and $\varphi = 0$
		Analytical	Numerical	
2	0.91136	0.91136	0.91136	0.911356
7	1.89540	1.89540	1.89540	1.89540

Graphical representations of the other numerical data for the variations of the related parameters of the present study are provided in Figures 3.18–3.20.

Figure 3.18a exhibits the velocity profiles for several values of the unsteadiness parameter A. It is seen that the velocity along the sheet initially decreases with the increase in the unsteadiness parameter, indicating an associated decrease in momentum boundary layer thickness. Later, the velocity of the Cu-water nanofluid rises with increasing A. With greater unsteadiness, there is more resistance flow near the surface and so fluid velocity initially decreases as the unsteady stretching velocity is larger near the surface. The velocity gradient at the surface $f''(0)$ diminishes with the increase in unsteadiness (Table 3.14). The highest value of the velocity gradient at the surface occurs with steady flow (i.e., for $A=0$) (Table 3.14).

Figure 3.18b and Figure 3.18c represent the effects of unsteadiness on the temperature and concentration of Cu-water nanofluid, respectively. From both the figures, it is observed that both the temperature and the concentration at a particular point are found to increase significantly with increasing unsteadiness. The temperature gradient at the surface rises with the rise in A for Cu-water nanofluid (Table 3.14). Figure 3.18c shows the rise in concentration for increasing values of the unsteadiness parameter. The concentration gradient at the surface also rises with the rise in unsteadiness. This finding agrees well with Figure 3.18c.

Figure 3.19a shows that the velocity of Cu-water nanofluid diminishes with the rise in nanoparticle volume fraction φ. The temperature of the Cu-water nanofluid rises with an increase in nanoparticle volume fraction φ, and it is evident as the rate of heat transfer at the surface reduces with a rise in φ as listed in Table 3.15. An increase in concentration of Cu-water nanofluid is noted with the rising values of φ as shown in Figure 3.19c.

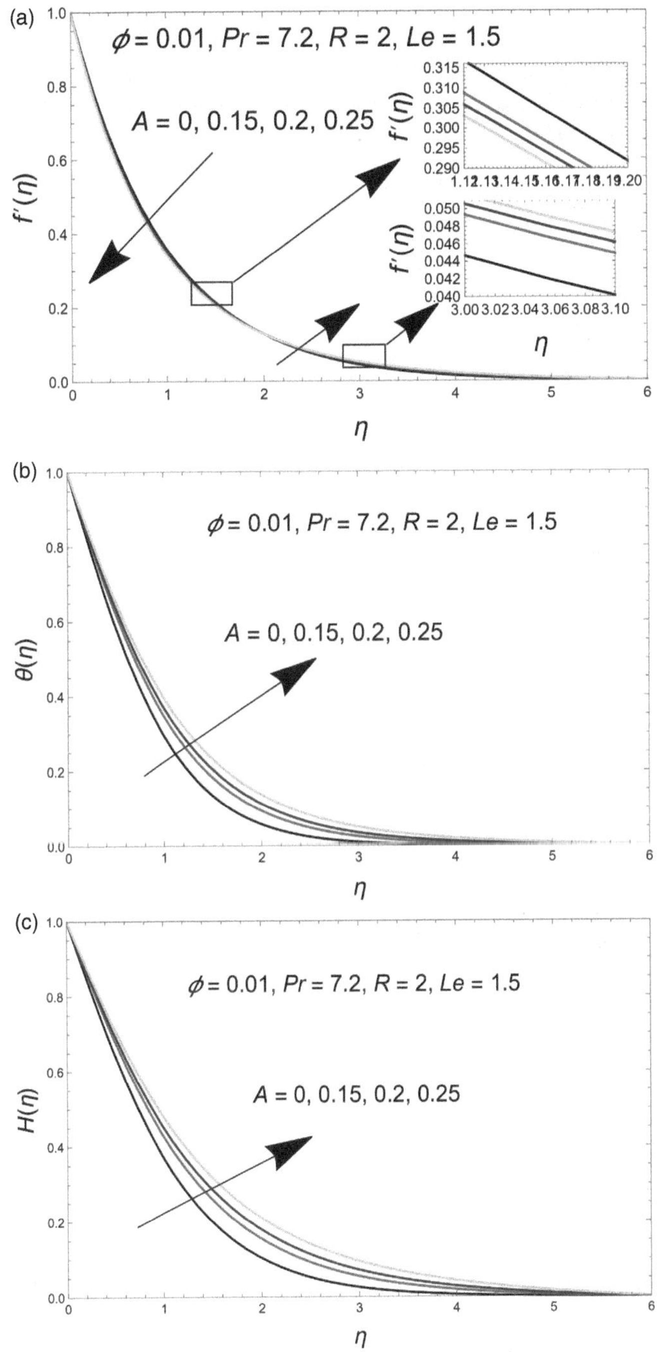

FIGURE 3.18 Effect of A on (a) velocity, (b) temperature, and (c) concentration profiles of Cu-water with $R = 2$, $Le = 1.5$, $\varphi = 0.01$, and $Pr = 7.2$.

TABLE 3.14
The value of $-f''(0)$, $-\theta'(0)$ and $-H'(0)$ for various values of parameter A with $Le = 1.5$, $Pr = 6.2$, $\varphi = 0.01$, $R = 2$

A	$-f''(0)$	$-\theta'(0)$	$-H'(0)$
0.0	1.02648	0.809588	0.754647
0.1	1.06151	0.757439	0.703655
0.2	1.09624	0.695792	0.643261
0.4	1.16443	0.510697	0.467658
0.5	1.19772	0.363088	0.338176

Figure 3.20 displays the rise in temperature of Cu-water nanofluid for a rise in thermal radiation parameter R. The temperature gradient at the surface rises with an increase in the radiation parameter R (Table 3.16).

Table 3.17 also indicates that the temperature gradient at the surface for Cu-water nanofluid diminishes with increases in Pr. Due to this imperfection, Pr is used to enhance the rate of cooling in nanofluid flows. A reduction in thermal boundary layer thickness is observed in this case (Nandeppanavar et al., 2011). As the Lewis number Le increases, the concentration gradient at the surface of Cu-water nano liquid is reduced (Table 3.18).

3.5 CONCLUSIONS

In this chapter, nanofluid flow, heat, and mass transfer past a stretched surface have been investigated in detail. Various types of stretching surface (e.g., linear stretching surface, nonlinear stretching surface, exponential stretching surface and unsteady stretching surface) have been considered. Both single-phase and two-phase models of nanofluid have been considered. The effects of Brownian motion and thermophoresis have been analyzed in detail. Special emphasis has been placed on studying the effects of thermal radiation on nanofluid flow past an unsteady stretching surface. The results obtained show that such types of problems have wider technological applications and deserve special attention in the future.

Nomenclature

A unsteadiness parameter
C nanoparticle concentration
C_w surface concentration
C_∞ free stream concentration
c_p specific heat (J/kg K)
D_B Brownian diffusion coefficient
D_T thermophoresis diffusion coefficient
Le Lewis number
Nb Brownian motion parameter

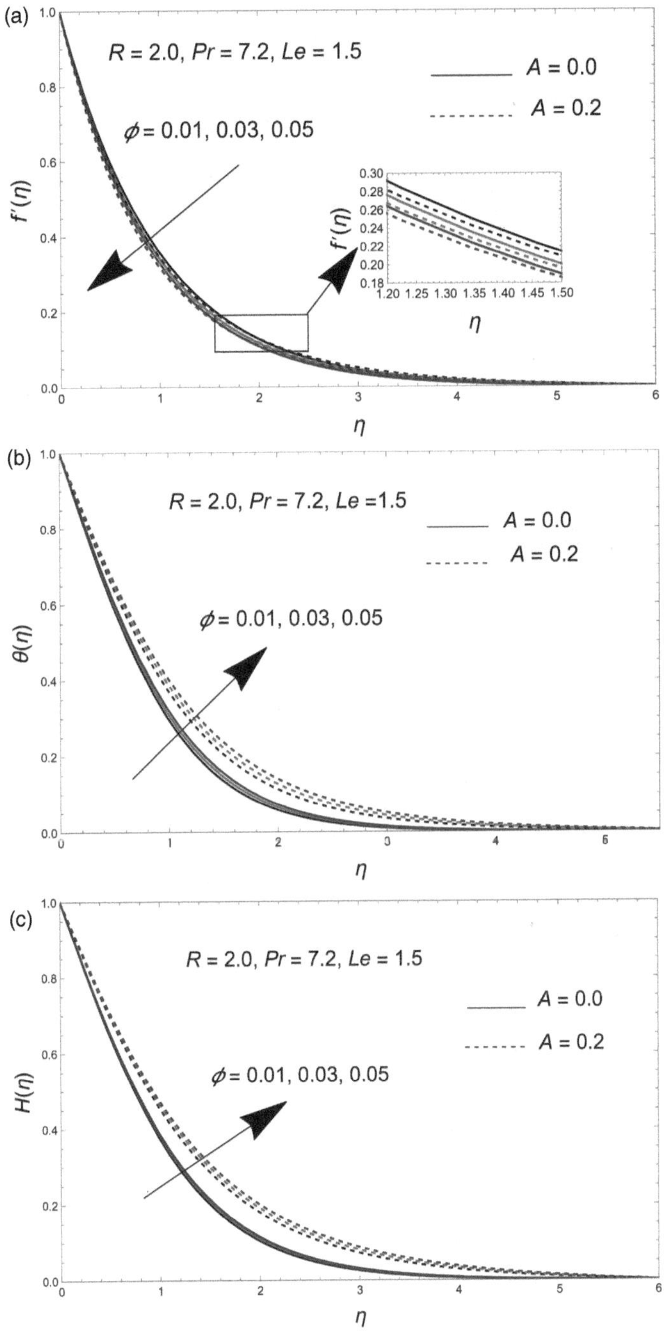

FIGURE 3.19 Effects of φ on (a) velocity, (b) temperature, and (c) concentration profiles of Cu-water with $R = 2$, Le =1.5, and $Pr = 7.2$.

TABLE 3.15
The value of –f'' (0), –θ' (0) and –H′(0) for various values of parameter φ fluid with Le = 1.5, Pr = 6.2, A = 0.2, R = 2

ϕ	$n - f''$ (0)	$-\theta'$ (0)	–H′(0)
0.01	1.09624	0.695792	0.643261
0.015	1.10920	0.689668	0.639677
0.02	1.12156	0.683706	0.636258
0.025	1.13336	0.677901	0.632996
0.03	1.14461	0.672249	0.629885
0.035	1.15535	0.666743	0.626920

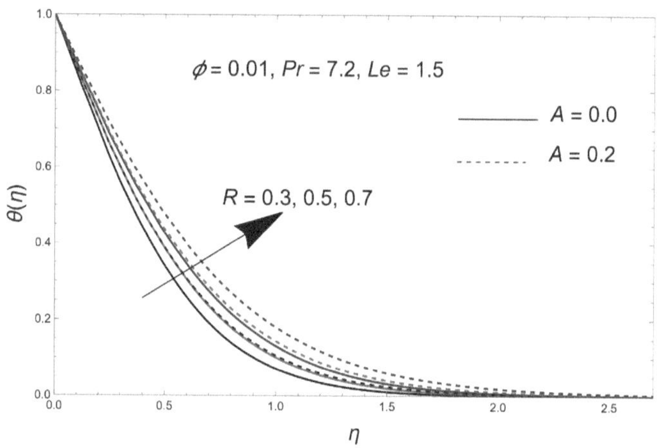

FIGURE 3.20 Effects of R on the temperature profile of Cu-water with Le = 1.5, φ = 0.01, and Pr = 7.2.

TABLE 3.16
The value of $-\theta'$ (0) for various values of parameter R with Le = 1.5, Pr = 6.2, A = 0.2, φ = 0.01

R	$-\theta'$ (0)
0.0	1.585310
0.5	1.171250
1.0	0.946421
1.5	0.799987
2.0	0.695792
2.5	0.617921
3.0	0.557858

TABLE 3.17
The value of $-\theta'(0)$ for various values of *Pr* with *Le* = 1.5, *R* = 2, *A* = 0.2, φ = 0.01

Pr	$-\theta'(0)$
0.5	0.202086
3.0	0.417212
4.0	0.508337
5.0	0.596379
6.0	0.679733

TABLE 3.18
The value of –H′ (0) for various values of parameter *Le* fluid with *Pr* = 6.2, *R* = 2, *A* = 0.2, φ = 0.01

Le	$-H'(0)$
0.1	0.192548
0.5	0.313217
0.7	0.379886
1.2	0.547936
2.0	0.788417

Nt thermophoresis parameter
Pr Prandtl number
R thermal radiation parameter
S suction/blowing parameter
T temperature of the liquid (K)
T_w surface temperature (K)
T_∞ free stream temperature (K)
u,v velocity component in x, y direction (m/s)
η similarity variable
κ liquid thermal conductivity ($Wm^{-1}K^{-1}$)
v kinematic viscosity (m^2/s)
φ non-dimensional concentration
ρ liquid density (kg m^{-3})
θ non-dimensional fluid temperature

REFERENCES

Ahmad, K. and Ishak, A., 2017. Magnetohydrodynamic (MHD) Jeffrey fluid over a stretching vertical surface in a porous medium. *Propulsion and Power Research*, *6*(4), pp.269–276.

Bisht, A. and Maheshwari, S., 2023. Magnetized Sisko nanofluid flow over nonlinear stretching sheet: A computational approach. *Numerical Heat Transfer, Part A: Applications*, pp.1–21. https://doi.org/10.1080/10407782.2023.2242613

Chamkha, A.J., Aly, A.M. and Mansour, M.A., 2010. Similarity solution for unsteady heat and mass transfer from a stretching surface embedded in a porous medium with suction/injection and chemical reaction effects. *Chemical Engineering Communications*, *197*(6), pp.846–858.

Dey, D. and Borah, R., 2023. MHD nanofluid flow with energy transfer over a porous stretching surface by using a second-grade fluid model. *Numerical Heat Transfer, Part A: Applications*, *86*, pp. 1313—1327. https://doi.org/10.1080/10407782.2023.2278715

Elanchezhian, E., Nirmalkumar, R., Balamurugan, M., Mohana, K., Prabu, K.M. and Viloria, A., 2020. Heat and mass transmission of an Oldroyd-B nanofluid flow through a stratified medium with the swimming of motile gyrotactic microorganisms and nanoparticles. *Journal of Thermal Analysis and Calorimetry*, *141*(6), pp.2613–2623.

Elbashbeshy, E.M.A., 2001. Heat transfer over an exponentially stretching continuous surface with suction. *Archives of Mechanics*, *53*(6), pp.643–651.

Gamar, F., Shamshuddin, M.D., Ram, M.S., Salawu, S.O. and Fatunmbi, E.O., in press. Exploration of thermal radiation and stagnation point in MHD micropolar nanofluid flow over a stretching sheet with Navier slip. *Numerical Heat Transfer, Part A: Applications.* https://doi.org/10.1080/10407782.2024.2338264

Ghosh, S. and Mukhopadhyay, S., 2017. MHD slip flow and heat transfer of Casson nanofluid over an exponentially stretching permeable sheet. *International Journal of Automotive and Mechanical Engineering*, *14*(4), pp.4785–4804.

Gupta, P.S. and Gupta, A.S., 1977. Heat and mass transfer on a stretching sheet with suction or blowing. *The Canadian Journal of Chemical Engineering*, *55*(6), pp.744–746.

Hamad, M.A.A., 2011. Analytical solution of natural convection flow of a nanofluid over a linearly stretching sheet in the presence of magnetic field. *International Communications in Heat and Mass Transfer*, *38*(4), pp.487–492.

Haroun, N.A., Mondal, S. and Sibanda, P., 2017. Effects of thermal radiation on mixed convection in a MHD nanofluid flow over a stretching sheet using a spectral relaxation method. *International Journal of Mathematical, Computational, Physical, Electrical and Computer Engineering*, *11*(2), pp.33–43.

Hayat, U. and Shahzad, A., 2023. Analysis of heat transfer and thin film flow of Au– Np over an unsteady radial stretching sheet. *Numerical Heat Transfer, Part A: Applications*, *84*(11), pp.1338–1351.

Hussain, A., Malik, M.Y., Khan, M. and Salahuddin, T., 2020. Application of generalized Fourier heat conduction law on MHD viscoinelastic fluid flow over stretching surface. *International Journal of Numerical Methods for Heat & Fluid Flow*, *30*(6), pp.3481–3496.

Ishfaq, N., Khan, Z.H., Khan, W.A. and Culham, R.J., 2016. Estimation of boundary-layer flow of a nanofluid past a stretching sheet: A revised model. *Journal of Hydrodynamics*, *28*(4), pp.596–602.

Kandasamy, R., Loganathan, P. and Arasu, P.P., 2011. Scaling group transformation for MHD boundary-layer flow of a nanofluid past a vertical stretching surface in the presence of suction/injection. *Nuclear Engineering and Design*, *241*(6), pp.2053–2059.

Kathyayani, G. and Prakash Gowd, P., 2023. Influence of various shape factors on a hybrid nanofluid flow over a convectively heated exponential stretching sheet with exponential heat source and activation energy: an irreversibility analysis. *Numerical Heat Transfer, Part A: Applications*, *86*, pp. 1395–1414. https://doi.org/10.1080/10407782.2023.2279257

Khader, M.M., Babatin, M.M. and Megahed, A.M., 2023. Non-Newtonian nanofluid flow across an exponentially stretching sheet with viscous dissipation: numerical study using an SCM based on Appell–Changhee polynomials. *Boundary Value Problems*, *2023*(77), pp.1–18, https://doi.org/10.1186/s13661-023-01765-8

Khan, M.I., Alzahrani, F., Hobiny, A. and Ali, Z., 2020. Modeling of Cattaneo-Christov double diffusions (CCDD) in Williamson nanomaterial slip flow subject to porous medium. *Journal of Materials Research and Technology*, *9*(3), pp.6172–6177.

Khan, W.A. and Pop, I., 2010. Boundary-layer flow of a nanofluid past a stretching sheet. *International Journal of Heat and Mass Transfer*, *53*(11–12), pp.2477–2483.

Konai, S. and Mukhopadhyay, S., 2024. Insight into the forced convective radiative Stefan flow of nanofluid over an unsteady stretched sheet. *International Journal of Computational Materials Science and Engineering*, pp.2450017. https://doi.org/10.1142/S2047684124500179

Konai, S., Maiti, H. and Mukhopadhyay, S., 2023. Influences of Stefan blowing on unsteady flow of Casson nanofluid past a stretching surface. *Forces in Mechanics*, *12*, pp.100227. https://doi.org/10.1016/j.finmec.2023.100227

Loganathan, K., Alessa, N., Namgyel, N. and Karthik, T.S., 2021. MHD flow of thermally radiative Maxwell fluid past a heated stretching sheet with Cattaneo–Christov dual diffusion. *Journal of Mathematics*, *2021*(1), pp.5562667.

Loganathan, K., Mohana, K., Mohanraj, M., Sakthivel, P. and Rajan, S., 2020a. Impact of third-grade nanofluid flow across a convective surface in the presence of inclined Lorentz force: An approach to entropy optimization. *Journal of Thermal Analysis and Calorimetry*, *144*(5), pp.1935–1947.

Loganathan, K., Muhiuddin, G., Alanazi, A.M., Alshammari, F.S., Alqurashi, B.M. and Rajan, S., 2020b. Entropy optimization of third-grade nanofluid slip flow embedded in a porous sheet with zero mass flux and a non-Fourier heat flux model. *Frontiers in Physics*, *8*, pp.250. https://doi.org/10.3389/fphy.2020.00250

Magyari, E. and Keller, B., 1999. Heat and mass transfer in the boundary layers on an exponentially stretching continuous surface. *Journal of Physics D: Applied Physics*, *32*(5), pp.577. https://doi.org/10.1088/0022-3727/32/5/012

Mahato, R., Das, M., Sen, S.S.S. and Shaw, S., 2022. Entropy generation on unsteady stagnation-point Casson nanofluid flow past a stretching sheet in a porous medium under the influence of an inclined magnetic field with homogeneous and heterogeneous reactions. *Heat Transfer*, *51*(6), pp.5723–5747.

Mahmood, Z. and Khan, U., 2023. Mixed convective flow of nanofluid across exponential surface: A numerical assessment of the impact of Darcy-Forchheimer and nanoparticle aggregation. *Numerical Heat Transfer, Part A: Applications*, *86*, pp. 2192–2217. https://doi.org/10.1080/10407782.2023.2288265

Mahmood, Z., Z. Duraihem, F., Adnan, Khan, U. and Hassan, A.M., 2024. Model-based comparative analysis of MHD stagnation point flow of hybrid nanofluid over a stretching sheet with suction and viscous dissipation. *Numerical Heat Transfer, Part B: Fundamentals*, pp.1–22. https://doi.org/10.1080/10407790.2024.2318457

Majeed, A., Zeeshan, A. and Ellahi, R., 2017. Chemical reaction and heat transfer on boundary layer Maxwell Ferro-fluid flow under magnetic dipole with Soret and suction effects. *Engineering Science and Technology, An International Journal*, *20*(3), pp.1122–1128.

Makkar, V., Hayat, T. and Alsaedi, A., 2021. A mathematical simulation of unsteady MHD Casson nanofluid flow subject to the influence of chemical reaction over a stretching surface: Buongiorno's model. *Heat Transfer*, *50*(8), pp.8640–8655.

Mandal, G. and Pal, D., 2023. Estimation of entropy generation and heat transfer of magnetohydrodynamic quadratic radiative Darcy–Forchheimer cross hybrid nanofluid (GO+ Ag/

kerosene oil) over a stretching sheet. *Numerical Heat Transfer, Part A: Applications*, *84*(8), pp.853–876.

Mandal, G. and Pal, D., 2024. Dual solutions of Darcy-Forchheimer radiative-convective magneto-nanofluid flow over an exponentially stretching/shrinking surface: Stability and entropy generation analysis. *Numerical Heat Transfer, Part A: Applications*, *86*, pp. 4601–4633. https://doi.org/10.1080/10407782.2024.2321515

Mandal, S. and Shit, G.C., 2021. Entropy analysis of unsteady MHD three-dimensional flow of Williamson nanofluid over a convectively heated stretching sheet. *Heat Transfer*, *51*(2), pp.2034–2062.

Mukhopadhyay, S., Mondal, I.C. and Gorla, R.S.R., 2012. Effects of thermal stratification on flow and heat transfer past a porous vertical stretching surface. *Heat and Mass Transfer*, *48*, pp.915–921.

Naidu, K.K., Babu, H.D., Reddy, H.S. and Narayana, S.P., 2024. Magneto-Stefan blow enhanced heat and mass transfer flow in non-Newtonian ternary hybrid nanofluid across the nonlinear elongated surface. *Numerical Heat Transfer Part B-Fundamentals*, *86*, pp.1463–1482. https://doi.org/10.1080/10407790.2024.2314133

Nandeppanavar, M.M., Vajravelu, K., Abel, M.S. and Ng, C.O., 2011. Heat transfer over a nonlinearly stretching sheet with non-uniform heat source and variable wall temperature. *International Journal of Heat and Mass Transfer*, *54*(23–24), pp.4960–4965.

Nandi, S. and Kumbhakar, B., 2023. Computational analysis of hybrid nanoliquid flow over an exponentially elongating sheet influenced by oblique Lorentz force with viscous-Joule dissipation: Role of nanoparticle shape factor. *Numerical Heat Transfer, Part A: Applications*, *86*, pp.996–1019. https://doi.org/10.1080/10407782.2023.2277822

Naseem, T. and Shahzad, A., 2023. Water-based nanofluid flow and heat transfer in thin film over an unsteady stretching sheet: An entropy analysis. *Numerical Heat Transfer, Part A: Applications*, *86*, pp.1201–1225. https://doi.org/10.1080/10407782.2023.2273453

Naveed Khan, M., Khan, A.A., M. Alqahtani, A., Wang, Z., Hejazi, H.A. and Az-Zo'bi, E.A., 2024. Chemically reactive aspects of stagnation-point boundary layer flow of second-grade nanofluid over an exponentially stretching surface. *Numerical Heat Transfer, Part B: Fundamentals*, *86*, pp.1622–1638. https://doi.org/10.1080/10407790.2024.2318456

Negi, S., Rawat, S.K. and Kumar, M., 2021. Cattaneo–Christov double-diffusion model with Stefan blowing effect on copper–water nanofluid flow over a stretching surface. *Heat Transfer*, *50*(6), pp.5485–5515.

Pal, D. and Mandal, G., 2014. Influence of thermal radiation on mixed convection heat and mass transfer stagnation-point flow in nanofluids over stretching/shrinking sheet in a porous medium with chemical reaction. *Nuclear Engineering and Design*, *273*, pp.644–652.

Pal, D., 2009. Heat and mass transfer in stagnation-point flow towards a stretching surface in the presence of buoyancy force and thermal radiation. *Meccanica*, *44*(2), pp.145–158.

Prasannakumara, B.C., Madhukesh, J.K. and Ramesh, G.K., 2023. Bioconvective nanofluid flow over an exponential stretched sheet with thermophoretic particle deposition. *Propulsion and Power Research*, *12*(2), pp.284–296.

Preetham, M.P. and Kumbinarasaiah, S., 2024. A numerical study of two-phase nanofluid MHD boundary layer flow with heat absorption or generation and chemical reaction over an exponentially stretching sheet by Haar wavelet method. *Numerical Heat Transfer, Part B: Fundamentals*, *85*(6), pp.706–735.

Ramanjini, V., Krishna, G.G., Mishra, S.R., Kumari, S.S. and Sree, H.K., 2022. An unsteady axisymmetric Williamson nanofluid flow over a radially stretching Riga plate for the inclusion of mixed convection and thermal radiation. *Partial Differential Equations in Applied Mathematics*, *6*, p.100456. https://doi.org/10.1016/j.padiff.2022.100456

Ramya, D., Raju, R.S., Rao, J.A. and Chamkha, A.J., 2018. Effects of velocity and thermal wall slip on magnetohydrodynamics (MHD) boundary layer viscous flow and heat transfer of a nanofluid over a non-linearly-stretching sheet: A numerical study. *Propulsion and Power Research*, *7*(2), pp.182–195.

Rashid, U., Ullah, N., Akgül, A., Lu, D. and Ul Rahman, J., 2024. Micropolar (copper–water) nanofluid flow past a stretching sheet with nanosized particles shape impact. *Numerical Heat Transfer, Part B: Fundamentals*, pp.1–14. https://doi.org/10.1080/10407790.2024.2337134

Rashidi, M.M., Ganesh, N.V., Hakeem, A.A. and Ganga, B., 2014. Buoyancy effect on MHD flow of nanofluid over a stretching sheet in the presence of thermal radiation. *Journal of Molecular Liquids*, *198*, pp.234–238.

Razzaq, R. and Farooq, U., 2024. Non-similar analysis of MHD hybrid nanofluid flow over an exponentially stretching/shrinking sheet with the influences of thermal radiation and viscous dissipation. *Numerical Heat Transfer, Part B: Fundamentals*, pp.1–16. https://doi.org/10.1080/10407790.2024.2312958

Reddy, Y.D. and Mangamma, I., 2023. Significance of radiation and chemical reaction on MHD heat transfer nanofluid flow over a nonlinearly porous stretching sheet with nonuniform heat source. *Numerical Heat Transfer, Part A: Applications*, *85*(18), pp.2940–2966.

Renuka, S., Narender, G. and Govardhan, K., 2023. The effect of a slip wall on radiate magnetohydrodynamic Casson nanofluid past a stretching surface with viscous dissipation: Numerical approach. *Numerical Heat Transfer, Part B: Fundamentals*, pp.1–15. https://doi.org/10.1080/10407790.2023.2274447

Sahoo, B. and Poncet, S., 2011. Flow and heat transfer of a third grade fluid past an exponentially stretching sheet with partial slip boundary condition. *International Journal of Heat and Mass Transfer*, *54*(23–24), pp.5010–5019.

Salahuddin, T., Malik, M.Y., Hussain, A., Bilal, S. and Awais, M., 2016. MHD flow of Cattanneo–Christov heat flux model for Williamson fluid over a stretching sheet with variable thickness: Using numerical approach. *Journal of Magnetism and Magnetic Materials*, *401*, pp.991–997.

Sandeep, N. and Sulochana, C., 2018. Momentum and heat transfer behaviour of Jeffrey, Maxwell and Oldroyd-B nanofluids past a stretching surface with non-uniform heat source/sink. *Ain Shams Engineering Journal*, *9*(4), pp.517–524.

Sankari, M.S., Rao, M.E., Khan, W., Makinde, O.D., Chamkha, A.J. and Salah, B., in press. Homotopic analysis for bioconvection of Casson nanofluid flow over an exponential stretching sheet with activation energy and motile microorganism. *Numerical Heat Transfer, Part A: Applications*. https://doi.org/10.1080/10407782.2024.2333041

Santosh, C. and Parida, S.K., 2023. Heat transfer analysis of MHD viscous nanofluid flow over a nonlinearly stretching sheet with heat source/sink: A numerical study. *Numerical Heat Transfer, Part A: Applications*, *86*, pp.462–481. https://doi.org/10.1080/10407782.2023.2266568

Sewvandi, P., Dewasurendra, M. and Vajravelu, K., 2024. Triple-diffusive magneto-nanofluid flow over a nonlinearly stretching sheet. *Numerical Heat Transfer, Part A: Applications*, *86*, pp.996–1019. https://doi.org/10.1080/10407782.2024.2331002

Sharidan, S., Mahmood, M. and Pop, I., 2006. Similarity solutions for the unsteady boundary layer flow and heat transfer due to a stretching sheet. *Applied Mechanics and Engineering*, *11*(3), pp.647–654.

Sulochana, C. and Mahalaxmi, B., 2022. Thermophoresis and Brownian motion effects on Williamson nanofluid flow past a stretching surface with thermal radiation and chemical reaction. *Heat Transfer*, *51*(3), pp.2761–2779.

Triveni, B., Rao, M.V.S., Gangadhar, K. and Chamkha, A.J., 2023. Heat transfer analysis of MHD Casson nanofluid flow over a nonlinear stretching sheet in the presence of nonuniform heat source. *Numerical Heat Transfer, Part A: Applications*, *85*(13), pp.2145–2164.

Turner, M.R. and Weidman, P., 2021. Homann stagnation-point flow impinging on a biaxially stretching surface. *European Journal of Mechanics-B/Fluids*, *86*, pp.49–56.

Vajravelu, K. and Mukhopadhyay, S., 2015. *Fluid flow, heat and mass transfer at bodies of different shapes: Numerical solutions*. New York: Academic Press.

Wang, C.Y., 1989. Free convection on a vertical stretching surface. *ZAMM-Journal of Applied Mathematics and Mechanics/Zeitschrift für Angewandte Mathematik und Mechanik*, *69*(11), pp.418–420.

Wang, C.Y., 1989. Free convection on a vertical stretching surface. *ZAMM-Journal of Applied Mathematics and Mechanics/Zeitschrift für Angewandte Mathematik und Mechanik*, *69*(11), pp.418–420.

Zeeshan, A. and Majeed, A., 2016. Heat transfer analysis of Jeffery fluid flow over a stretching sheet with suction/injection and magnetic dipole effect. *Alexandria Engineering Journal*, *55*(3), pp.2171–2181.

4 Nanofluid Flow at a Stretching Surface

4.1 INTRODUCTION

A mixture of two different types of nanoparticles in a base fluid gives rise to what is termed a *hybrid nanofluid*. Recently, hybrid nanofluids have been a center of attention for researchers to enhance the thermal efficiency of heat exchangers, among other applications. The effect of a hybrid nanofluid flow on a rotating permeable stretching sheet and found that the hybrid nanofluid provides higher thermal efficiency than the base fluid was studied by Arshad and Hassen (2022). Mohd et al. (2022) investigated unsteady three-dimensional flow in a rotating hybrid nanofluid over a stretching sheet, where they used different parameter effects to analyze the hybrid nanofluid flow. Jamaludin et al. (2023) studied opposing MHD flow of $Cu\text{-}TiO_2$ hybrid nanofluid embedded in a porous medium with a heat source and slipped over an exponentially stretching/shrinking surface. An enhancement of the thermal performance of ethylene glycol by hybrid nanoparticles was examined by Nawaz et al. (2020).

Several studies have investigated the impact of magnetic fields on fluid flow and heat transfer. Bashir et al. (2024) examined the impact of nonlinear thermal radiation on the magnetized dissipative flow of a ternary hybrid nanomaterial. They found that magnetic fields significantly affect the flow behavior of a ternary hybrid nanofluid and observed that thermal radiation enhances heat transfer in the flow. Amin et al. (2023) studied the flow features and heat/mass transfer in a two-layer vertical channel with a $Gr\text{-}TiO_2$ hybrid nanofluid under MHD and radiation effects. They found that magnetic fields and radiation significantly influence the flow behavior and heat transfer in the channel. They observed that the hybrid nanofluid exhibits enhanced heat transfer compared to traditional fluids. Khan et al. (2024) investigated the effects of magnetic fields on the flow of Maxwell nanofluids with chemical reaction and thermal radiation on a stretching sheet. They showed that both chemical reactions and thermal radiation enhance heat transfer in the flow.

Porous media are crucial in various industries, including water reservoir management and water purification. Researchers have recently explored the complexities of fluid flow through porous media. For instance, Khan et al. (2019) investigated unsteady magneto-hydrodynamics (MHD) flow between two orthogonally moving porous plates. Their study revealed the flow characteristics and the impact of porous plates on fluid flow. Mishra and Kumar (2019) examined the influence of viscous

DOI: 10.1201/9781003571278-4

dissipation and Joule heating on fluid flow through a porous medium with thermal radiation. They used silver-water and copper-water nanofluids over a stretching sheet in their analysis. More recently, Anwar et al. (2024) investigated the effects of double stratification on the MHD flow of micropolar nanofluid over a stretching surface embedded in a porous medium. These studies demonstrate the significance of porous media in various engineering applications and highlight the need for continued research in this area.

Buoyancy is the upward force exerted by a fluid (such as water or air) on an object partially or fully immersed in the fluid, provided the density of the object differs from the fluid density. Waini et al. (2021) investigated the flow of a hybrid nanofluid towards a stagnation point on an exponentially stretching/shrinking sheet, taking into account buoyancy effects. This study examined the complex interactions between the hybrid nanofluid, the stretching/shrinking sheet, and the buoyancy forces. The study's findings can enhance heat transfer in various applications, such as electronics cooling and heat exchangers. A recent numerical study by Rehman et al. (2021) investigated the effects of buoyancy and radiation on magnetohydrodynamic (MHD) micropolar nanofluid flow over a stretching/shrinking sheet with a heat source. The study used a micropolar nanofluid, which exhibits micro-rotational effects, in the presence of a magnetic field. The sheet's motion created a complex flow pattern, with both stretching and shrinking regions. The study accounted for the buoyancy forces arising from density differences between the nanofluid and the surrounding fluid.

Thermal stratification refers to the layering of fluids due to temperature differences, resulting in distinct thermal layers. Concentration stratification is a crucial phenomenon resulting from density differences. Hayat et al. (2014) examined thermal and concentration effects in the mixed convective and radiative flow of a Jeffrey fluid past a stretching sheet. Shamshuddin et al. (2023) conducted a theoretical study of the convective flow of a Buongiorno nanofluid over a stretching Riga plate with dual stratification effects. Gul et al. (2020) investigated the flow of a 3D hybrid nanofluid with Hall current and ion slip over a nonlinear stretching surface in a thermally stratified medium. Masood and Farooq (2021) examined the behavior of graphene oxide and Silver nanoparticles immersed in water-based fluid over a stretching sheet with stratification and radiation effects.

Chemical reactions significantly impact fluid dynamics, particularly in applications involving reacting flows. The interplay between chemical reactions and fluid dynamics is essential in various fields. Mandal and Pal (2024) investigated the effects of binary chemical reactions with activation energy on magnetohydrodynamic (MHD) thermally radiating blood-based hybrid nanofluid flow through a shrinking stenotic artery. Their study focused on entropy generation and stability analysis. The activation energy affects the binary chemical reaction, which in turn impacts the flow behavior and heat transfer of the fluid. Madiwal and Naduvinamani (2024) explored the heat and mass transformation of a Casson hybrid nanofluid. This fluid is composed of MoS_2 and ZnO nanoparticles suspended in engine oil. This study contributes to the understanding of heat and mass transformation in Casson hybrid nanofluids. The findings have implications for various engineering applications, including heat transfer enhancement, mass transfer applications, and engine oil applications.

Based on the aforementioned literature, it is evident that there is research regarding the analysis of heat and mass transport of MHD flow of hybrid nanofluids over a stretching plate with viscous dissipation. The current study explores the mixed convective flow of titanium oxide and silicon oxide nanoparticles water as a base fluid over a stretching plate in the presence of a magnetic field, dual stratification, viscous dissipation, and chemical reactions. Obtaining a system of nonlinear ODEs is accomplished by similarity transformations. The bvp5c package from MATLAB® is utilized to obtain the numerical results. This program has a wide range of applications in the engineering and technology field, such as solar thermal energy storage, where it is used to optimize solar heat energy storage and improve the efficiency of solar water heating systems and building HVAC systems (Heating, Ventilation, and Air Conditioning systems).

4.2 MATHEMATICAL FORMULATION

Consider a two-dimensional (2D) steady, laminar, incompressible boundary layer flow of a hybrid nanofluid flow caused by a vertical stretching plate with a porous medium. In addition, a magnetic field, Joule heating, buoyancy, viscous dissipation, thermal stratification, and chemical reactions are considered. Figure 4.1 illustrates the schematic flow setup for the current flow model. Cartesian coordinate directions are x and y, with u and v representing the velocity components in these flow directions. The sheet's surface expands with uniform stretching velocity, $u_w = cx$, where c represents the stretching rate factor. Additionally, the temperature and concentration near the wall are expressed by T_w and C_w, where $T_w = T_0 + bx$ and $C_w = C_0 + ex$ and T_∞ and C_∞ represent the ambient temperature and concentration of the hybrid nanofluid respectively and are given by $T_\infty = T_0 + ax$ and $C_\infty = C_0 + dx$ where a, b, d, and e

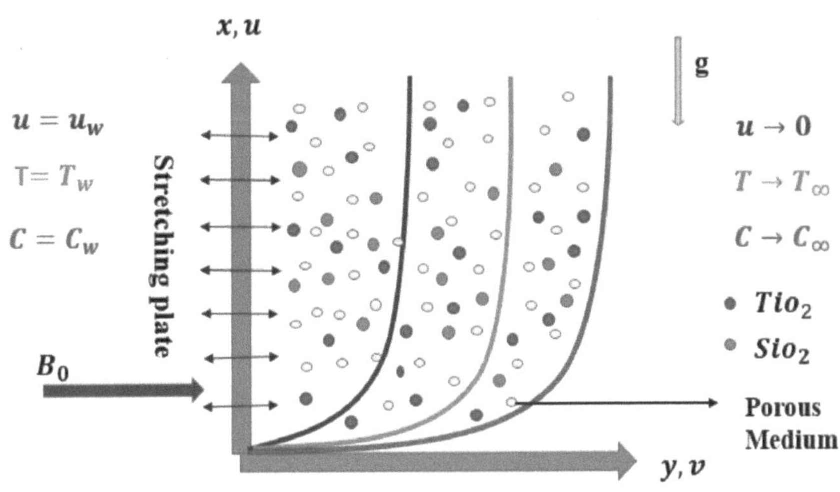

FIGURE 4.1 Flow geometry.

are dimensional constants and T_0 and C_0 are the reference temperature and reference concentration, respectively. Based on this discussion, the boundary layer equations are obtained as in references Hayat et al. (2014), Shamshuddin et al. (2023), Masood and Farooq (2021), Junaid et al. (2024), and Jamshed et al. (2021).

Conservation of mass:

$$\left(\frac{\partial u}{\partial x}\right)+\left(\frac{\partial v}{\partial y}\right)=0. \tag{4.1}$$

Momentum conservation:

$$u\frac{\partial u}{\partial x}+v\frac{\partial u}{\partial y}=\frac{\mu_{hnf}}{\rho_{hnf}}\frac{\partial^2 u}{\partial y^2}-\frac{\sigma_{hnf}}{\rho_{hnf}}B_0{}^2u-\left(\frac{\mu_{hnf}}{\rho_{hnf}}\right)\frac{u}{k^*}$$
$$+\frac{\left(\rho\beta_T\right)_{hnf}}{\rho_{hnf}}g\left(T-T_\infty\right)+\frac{\left(\rho\beta_C\right)_{hnf}}{\rho_{hnf}}g\left(C-C_\infty\right). \tag{4.2}$$

Conservation of energy:

$$u\frac{\partial T}{\partial x}+v\frac{\partial T}{\partial y}=\frac{k_{hnf}}{\left(\rho C_p\right)_{hnf}}\frac{\partial^2 T}{\partial y^2}+\frac{\mu_{hnf}}{\left(\rho C_p\right)_{hnf}}\left(\frac{\partial u}{\partial y}\right)^2+\frac{\sigma_{hnf}}{\left(\rho C_p\right)_{hnf}}Bo^2u^2. \tag{4.3}$$

Species conservation:

$$u\frac{\partial C}{\partial x}+v\frac{\partial C}{\partial y}=D_{hnf}\left(\frac{\partial^2 C}{\partial y^2}\right)-K_1(C-C_\infty). \tag{4.4}$$

These governing equations are subject to the corresponding wall and free stream boundary conditions as discussed in Hayat et al., 2014 and Masood and Farooq, 2021 reproduced here.

$$\left.\begin{array}{ll} u=u_w, v=0, \mathrm{T}=\mathrm{T}_w=\mathrm{T}_0+\mathrm{bx}, \mathrm{C}=\mathrm{C}_w=C_0+ex & at\ y=0. \\ u\rightarrow 0, T=T_\infty=\mathrm{T}_0+a\,\mathrm{x}, C=C_\infty=C_0+dx & as\ y\rightarrow\infty, \end{array}\right\} \tag{4.5}$$

where the dynamic viscosity is (μ_{hnf}), the density is (ρ_{hnf}), electrical conductivity is (σ_{hnf}), thermal expansion coefficient is (β_{Thnf}), solutal expansion coefficient is

(β_{Chnf}), thermal conductivity is (k_{hnf}), the heat capacitance (ρCp_{hnf}), and mass diffusivity is (D_{hnf}). These just-referenced thermophysical properties are calculated as follows:

$$\mu_{hnf} = \frac{\mu_f}{(1-\phi_1)^{2.5}(1-\phi_2)^{2.5}},$$

$$\rho_{hnf} = \phi_1\rho_{s1} + \phi_2\rho_{s2} + (1-\phi_1-\phi_2)\rho_f,$$

$$\sigma_{hnf} = \left(1 + \frac{3\left(\left(\frac{\sigma_{s2}}{\sigma_{nf}}\right)-1\right)\phi_2}{\left(\frac{\sigma_{s2}}{\sigma_{nf}}\right)+2-\left(\left(\frac{\sigma_{s2}}{\sigma_{nf}}\right)-1\right)\phi_2}\right)\sigma_{nf},$$

$$\sigma_{nf} = \left(1 + \frac{3\left(\left(\frac{\sigma_{s1}}{\sigma_f}\right)-1\right)\phi_1}{\left(\frac{\sigma_{s1}}{\sigma_f}\right)+2-\left(\left(\frac{\sigma_{s1}}{\sigma_f}\right)-1\right)\phi_1}\right)\sigma_f,$$

$$(\rho\beta)_{hnf} = (1-\phi_1-\phi_2)(\rho\beta)_f + \phi_1(\rho\beta)_{s1} + \phi_2(\rho\beta)_{s2},$$

$$K_{hnf} = \left(\frac{K_{s2}+2K_{nf}-2\phi_2\left(K_{nf}-K_{s2}\right)}{K_{s2}+2K_{nf}+\phi_2\left(K_{nf}-K_{s2}\right)}\right)K_{nf},$$

$$K_{nf} = \left(\frac{K_{s1}+2K_f-2\phi_1\left(K_f-K_{s1}\right)}{K_{s1}+2K_f+\phi_1\left(K_f-K_{s1}\right)}\right)K_{f},$$

$$\left(\rho C_p\right)_{hnf} = (1-\phi_1-\phi_2)\left(\rho C_p\right)_f + \phi_1\left(\rho C_p\right)_{s1} + \phi_2\left(\rho C_p\right)_{s2},$$

$$D_{hnf} = \frac{1}{(1-\phi_1)^{2.5}(1-\phi_2)^{2.5}}. \tag{4.6}$$

Next, the similarity transformation is initiated through the following relationships:

$$u = cxf', \; v = -(c\,\upsilon)^{1/2} f, \; \eta = \left(\frac{c}{\upsilon}\right)^{1/2} y,$$
$$\theta(\eta) = \frac{T - T_\infty}{T_w - T_0}, \; \phi(\eta) = \frac{C - C_\infty}{C_w - C_0}. \tag{4.7}$$

Utilization of Eq. (4.6) and similarity variables from Eq. (4.7) into Equations (4.1)–(4.4) along with boundary condition (4.5), yields:

$$M_1 f''' + ff'' - f'^2 - M_2 Mf' - M_1 Kf' + M_3 \lambda_1 \theta + M_3 \lambda_2 \phi = 0, \tag{4.8}$$

$$M_4 \theta'' - \Pr f'\theta - \Pr \delta_1 f' + \Pr f\theta' + \Pr M_5 Ecf''^2 + \Pr M_6 \, \mathrm{MEc}\, f'^2 = 0, \tag{4.9}$$

$$M_7 \phi'' + Scf\phi' - Scf'\phi - Scf'\delta_2 - ScKr\,\phi = 0. \tag{4.10}$$

The transformed boundary constraints are

$$\left.\begin{aligned} &f(0) = 0,\; f'(0) = 1,\; \theta(0) = 1 - \delta_1,\; \phi(0) = 1 - \delta_2, \\ &f'(\infty) \to 0,\; \theta(\infty) \to 0,\; \phi(\infty) \to 0. \end{aligned}\right\} \tag{4.11}$$

In the above expression

$$M_1 = \frac{1/(1-\phi_1)^{2.5}(1-\phi_2)^{2.5}}{\phi_1 \frac{\rho_{s1}}{\rho_f} + \phi_2 \frac{\rho_{s2}}{\rho_f} + (1 - \phi_1 - \phi_2)}, M_2 = \frac{N_1 N_2}{\phi_1 \frac{\rho_{s1}}{\rho_f} + \phi_2 \frac{\rho_{s2}}{\rho_f} + (1 - \phi_1 - \phi_2)},$$

$$M_{3T} = \frac{\phi_1 \frac{(\rho\beta_T)_{s1}}{(\rho\beta_T)_f} + \phi_2 \frac{(\rho\beta_T)_{s2}}{(\rho\beta_T)_f} + (1 - \phi_1 - \phi_2)}{\phi_1 \frac{\rho_{s1}}{\rho_f} + \phi_2 \frac{\rho_{s2}}{\rho_f} + (1 - \phi_1 - \phi_2)},$$

$$M_{3c} = \frac{\phi_1 \frac{(\rho\beta_c)_{s1}}{(\rho\beta_c)_f} + \phi_2 \frac{(\rho\beta_c)_{s2}}{(\rho\beta_c)_f} + (1 - \phi_1 - \phi_2)}{\phi_1 \frac{\rho_{s1}}{\rho_f} + \phi_2 \frac{\rho_{s2}}{\rho_f} + (1 - \phi_1 - \phi_2)},$$

$$M_4 = \frac{N_3 N_4}{\left(1-\phi_1-\phi_2\right)+\phi_1\dfrac{\left(\rho C_p\right)_{s1}}{\left(\rho C_p\right)_f}+\phi_2\dfrac{\left(\rho C_p\right)_{s2}}{\left(\rho C_p\right)_f}},$$

$$M_5 = \frac{1/\left(1-\phi_1\right)^{2.5}\left(1-\phi_2\right)^{2.5}}{\left(1-\phi_1-\phi_2\right)+\phi_1\dfrac{\left(\rho C_p\right)_{s1}}{\left(\rho C_p\right)_f}+\phi_2\dfrac{\left(\rho C_p\right)_{s2}}{\left(\rho C_p\right)_f}},$$

$$M_6 = \frac{N_1 N_2}{\left(1-\phi_1-\phi_2\right)+\phi_1\dfrac{\left(\rho C_p\right)_{s1}}{\left(\rho C_p\right)_f}+\phi_2\dfrac{\left(\rho C_p\right)_{s2}}{\left(\rho C_p\right)_f}},$$

$$M_7 = \frac{1}{\left(1-\phi_1\right)^{2.5}\left(1-\phi_2\right)^{2.5}},$$

$$N_1 = 1+\left(\frac{3\left(\left(\dfrac{\sigma_{s2}}{\sigma_{nf}}\right)-1\right)\phi_2}{\left(\dfrac{\sigma_{s2}}{\sigma_{nf}}\right)+2-\left(\left(\dfrac{\sigma_{s2}}{\sigma_{nf}}\right)-1\right)\phi_2}\right), N_2 = 1+\left(\frac{3\left(\left(\dfrac{\sigma_{s1}}{\sigma_f}\right)-1\right)\phi_1}{\left(\dfrac{\sigma_{s1}}{\sigma_f}\right)+2-\left(\left(\dfrac{\sigma_{s1}}{\sigma_f}\right)-1\right)\phi_1}\right),$$

$$N_3 = \frac{K_{s2}+2K_{nf}-2\phi_2\left(K_{nf}-K_{s2}\right)}{K_{s2}+2K_{nf}+\phi_2\left(K_{nf}-K_{s2}\right)}, N_4 = \frac{K_{s1}+2K_f-2\phi_1\left(K_f-K_{s1}\right)}{K_{s1}+2K_f+\phi_1\left(K_f-K_{s1}\right)}, \tag{4.12}$$

and the non-dimensional parameters are given below:

$$M = \frac{\sigma_f B_0^{\,2}}{\rho_f c},\ K = \frac{\mu_f}{\rho_f c k^*},\ Gr_T = \frac{g(\beta_T)_f(T_w - T_0)x^3}{\upsilon^2},$$

$$\lambda_1 = \frac{Gr_T}{\mathrm{Re}_x^2},\ Gr_c = \frac{g(\beta_c)_f(C_w - C_0)x^3}{\upsilon^2},\ \lambda_2 = \frac{Gr_C}{\mathrm{Re}_x^2},$$

$$\Pr = \frac{\upsilon_f}{\alpha_f}, \ Ec = \frac{c^2 x^2}{\left(C_p\right)_f \left(T_w - T_0\right)},$$

$$\delta_1 = \frac{a}{b},$$

$$\delta_2 = \frac{d}{e}. \tag{4.13}$$

Quantities like skin friction Cfx, Nusselt number Nu_x, and Sherwood number Sh_x are defined as in Shamshuddin et al. (2023), Gul et al. (2020), and Masood and Farooq (2021) and are reproduced here.

$$Cf_x = \frac{\tau_w}{\rho_f u_w^2}, Nu = \frac{x q_w}{K_f \left(T_w - T_\infty\right)}, Sh = \frac{x q_m}{D_f \left(C_w - C_\infty\right)}. \tag{4.14}$$

Here, τ_w, q_w, and q_m are the wall shear stress, wall heat flux, and surface mass flux, which are given by

$$\tau_w = \mu_{hnf}\left(\frac{\partial u}{\partial y}\right)_{y=0}, q_w = -K_{hnf}\left(\frac{\partial T}{\partial y}\right)_{y=0}, q_m = -D_{hnf}\left(\frac{\partial C}{\partial y}\right)_{y=0}. \tag{4.15}$$

Inserting Equations (4.6) and (4.7) in Equations (4.14) and (4.15), we obtain the following expression:

$$\sqrt{\mathrm{Re}_x}\, Cf_x = M_1 f''(0), \tag{4.16}$$

$$\mathrm{Re}_x^{-\frac{1}{2}}\, Nu = -N_3 N_4 \left(\frac{1}{1-\delta_1}\right)\theta'(0), \tag{4.17}$$

$$\mathrm{Re}_x^{-\frac{1}{2}}\, Sh = -M_7 \left(\frac{1}{1-\delta_2}\right)\phi'(0), \tag{4.18}$$

where $Re_x = \frac{x u_w}{\upsilon_f}$ is the Reynold's number.

4.3 SOLUTION METHODOLOGY

The MATLAB® solver bvp4c is used to solve the system of nonlinear ordinary differential equations (ODEs) with transformed boundary conditions. To apply this method, the ODEs are first converted into a system of first-order ODEs by introducing new variables.

$$\begin{Bmatrix} f, f', f'' \\ \theta, \theta' \\ \phi, \phi' \end{Bmatrix} = \begin{Bmatrix} f_1, f_2, f_3 \\ f_4, f_5 \\ f_6, f_7 \end{Bmatrix} \tag{4.19}$$

Then the equations take the form as:

$$\frac{1}{M_1}\left(f_2^2 - f_1 f_3 + M_2 M f_2 + M_1 K f_2 - M_{3T}\lambda_1 f_4 - M_{3c}\lambda_2 f_6\right) = 0, \tag{4.20}$$

$$\frac{1}{M_4}\left(Pr f_2 f_4 + Pr\delta_1 f_2 - Pr f_1 f_5 - Pr M_5 Ecf_3^2 - Pr M_6 MEcf_2^2\right) = 0, \tag{4.21}$$

$$\frac{1}{M_7}\left(Scf_2 f_6 + Scf_2\delta_2 - Scf_1 f_7 + ScKrf_6\right) = 0. \tag{4.22}$$

The transformed boundary conditions become:

At $\eta = 0$,

$$f_1 = 0,\ f_2 = 1,\ f_4 = 1 - \delta_1, f_6 = 1 - \delta_2. \tag{4.23}$$

As $\eta = \infty$

$$f_2 = 0,\ f_4 = 0,\ f_6 = 0. \tag{4.24}$$

The thermophysical properties of the nanofluid components are listed in Table 4.1.

4.4 RESULTS AND DISCUSSION

Here, solutions are represented graphically. The effect of each parameter affecting the flow is discussed in connection with the following figures.

The variation in velocity due to the magnetic parameter M is represented in Figure 4.2. As we increase the magnetic parameter, the fluid particles experience a centripetal force known as the Lorentz force which acts opposite to the direction of the fluid flow, thus reducing the velocity of the hybrid nanofluid. When we increase the porous parameter, the fluid particles experience more void spaces which act as

TABLE 4.1
Thermophysical properties for nanoparticles of the hybrid nanofluid

	Water	TiO_2	SiO_2
$C_p\left(\frac{J}{kg\,K}\right)$	4179	686.2	745
$\rho\left(\frac{kg}{m^3}\right)$	997.1	4250	2200
$k\left(\frac{W}{m\,K}\right)$	0.613	8.9538	1.4
$\beta\left(\frac{1}{K}\right)$	21×10^{-5}	0.9×10^{-5}	42.7×10^{-5}
$\sigma\left(\frac{S}{m}\right)$	5×10^{-6}	2.4×10^{6}	1×10^{-12}

Source: From Bashir et al. (2024) and Khan et al. (2020).

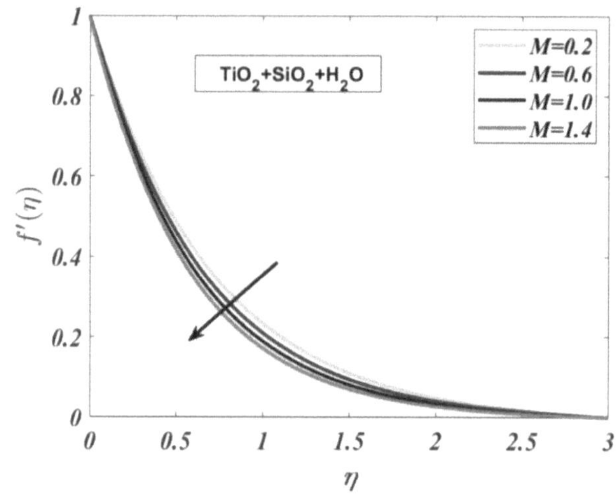

FIGURE 4.2 Change in $f'(\eta)$ due to M.

obstacles and result in a low velocity of hybrid nanofluid shown in Figure 4.3. The change in velocity due to the buoyancy parameter is represented in Figure 4.4 as the buoyancy parameter grows the convection in fluid gradually increases which results in a high velocity of hybrid nanofluid. The velocity profile grows with the growing solutal expansion parameter, resulting in forming a buoyancy force that drives the fluid flow, as can be seen in Figure 4.5.

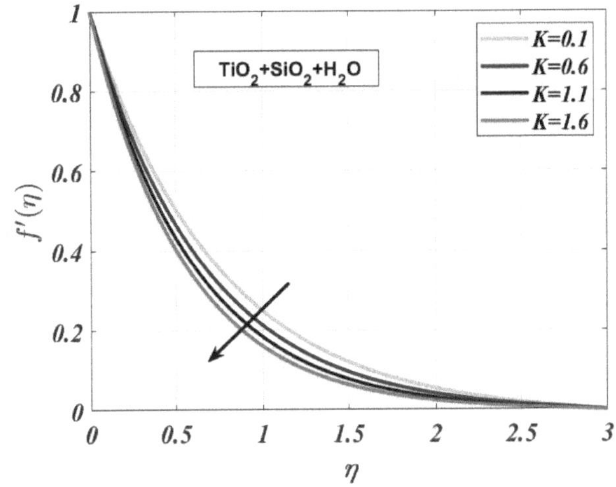

FIGURE 4.3 Change in $f'(\eta)$ due to K.

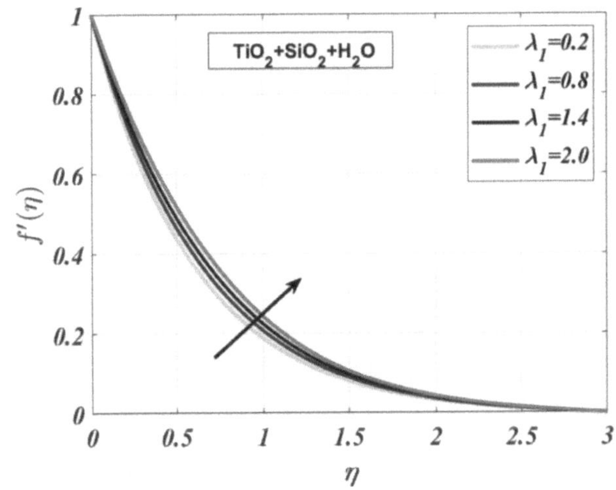

FIGURE 4.4 Change in $f'(\eta)$ due to λ_1.

Figure 4.6 depicts the temperature growing gradually due to the magnetic parameter. As the magnetic parameter increases, it gives rise to higher resistivity, which, in turn, increases the temperature of the hybrid nanofluid. The temperature profile rises due to the increase in the porous parameter. When there is an increase in the porous parameter, the resistance to the flow of the hybrid nanofluid also increases, which tends to increase the temperature as displayed in Figure 4.7. The thermal stratification parameter denoted by δ_1 decreases the temperature profile on increasing the thermal stratification parameter because it creates a stable density gradient that resists vertical mixing of the nanofluid which also prevents the warmer fluid from rising and the

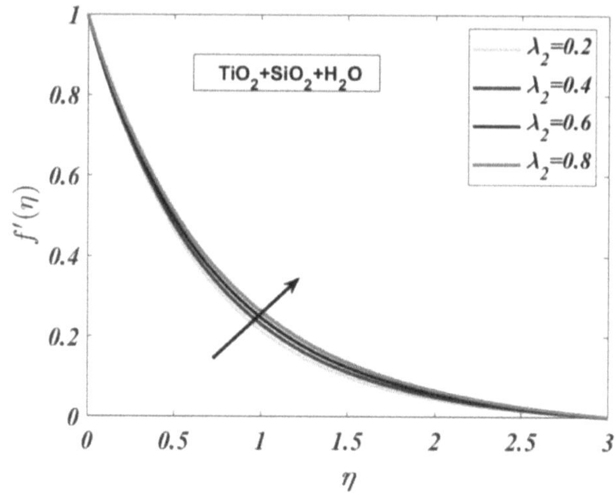

FIGURE 4.5 Change in $f'(\eta)$ due to λ_2.

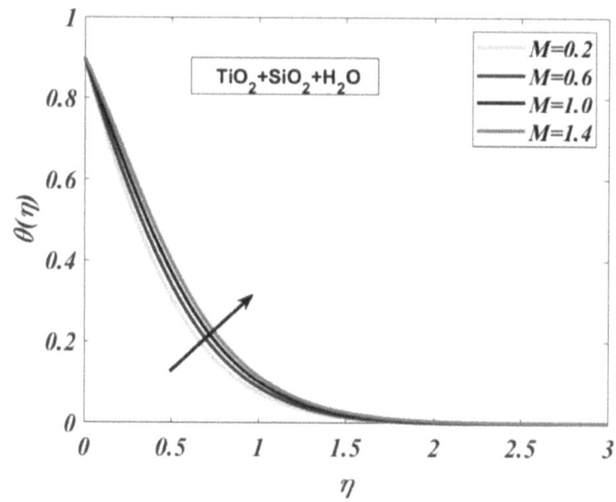

FIGURE 4.6 Change in $\theta(\eta)$ due to M.

cooler fluid from sinking, maintaining the temperature difference between the layers as shown in Figure 4.8. An increase in the Eckert number tends to increase the temperature as well because Eckert number quantifies the significance of viscous dissipation when we increase the Eckert number viscous dissipation increases which lead to increased internal energy by increasing the temperature of the hybrid nanofluid as displayed in Figure 4.9.

A higher solutal stratification creates a steeper concentration gradient between the surface and the ambient fluid. This gradient acts as a resistance to the diffusion of the nanoparticles into the surface of the fluid. It reduces the concentration profile

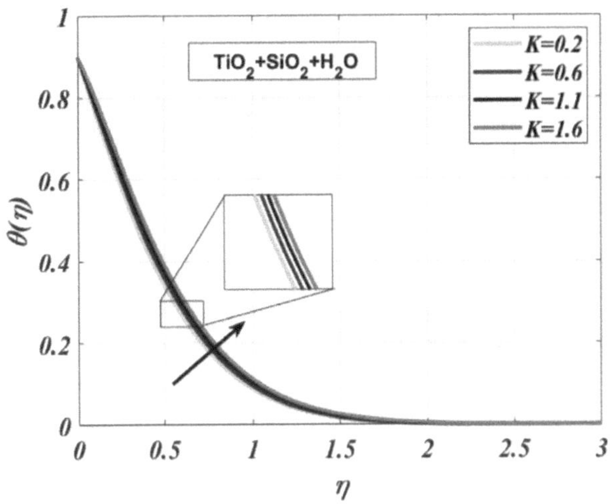

FIGURE 4.7 Change in $\theta(\eta)$ due to K.

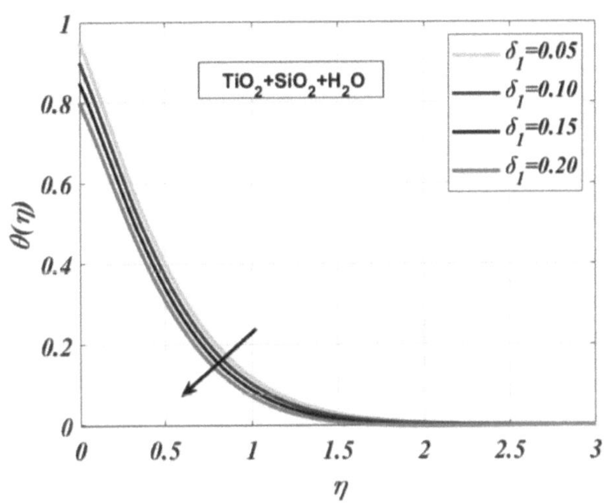

FIGURE 4.8 Change in $\theta(\eta)$ due to δ_1.

depicted in Figure 4.10. As we increase the chemical reaction parameter of the hybrid nanofluid, we observe that the concentration profile tends to decrease. This happens due to thinner boundary layer concentration, and lower peak concentration affects the concentration profile in Figure 4.11.

The skin friction due to the porosity parameter tends to decrease as the porosity parameter increases. As the porosity parameter increases and resistance decreases, fluid velocity can increase because there's less obstruction to the flow. However, the skin friction increases due to the buoyancy parameter; when there is an increase in the

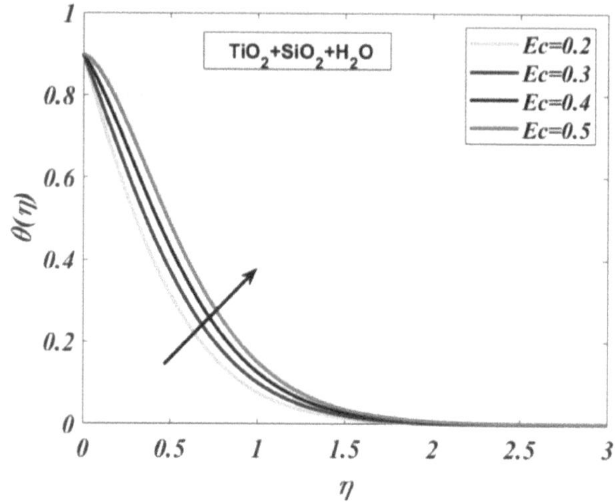

FIGURE 4.9 Change in $\theta(\eta)$ due to Ec.

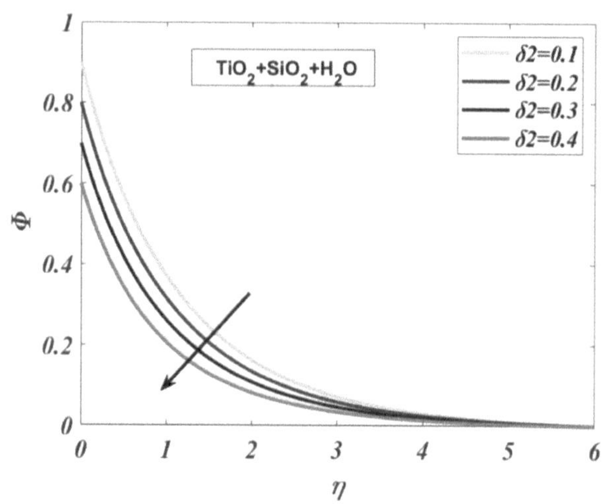

FIGURE 4.10 Change in $\Phi(\eta)$ due to δ_2.

buoyancy parameter, the velocity also increases, causing increased skin friction, as shown in Figure 4.12. We observe a decrease in the heat transfer profile in Figure 4.13 due to an increase in the thermal stratification parameter and Eckert number, as discussed in Figure 4.8 and Figure 4.9, respectively. The mass transfer coefficient increases near the walls of the fluid due to the chemical reaction parameter and solutal stratification parameter, as the effect of these parameters is higher near the walls than the surface, as shown in Figure 4.14 (Table 4.2).

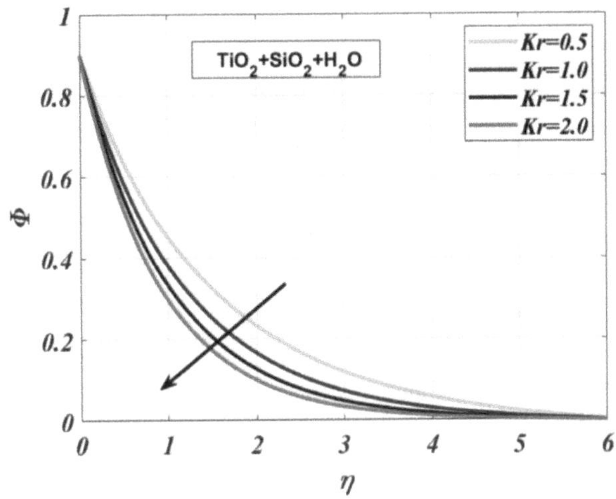

FIGURE 4.11 Change in $\Phi(\eta)$ due to Kr.

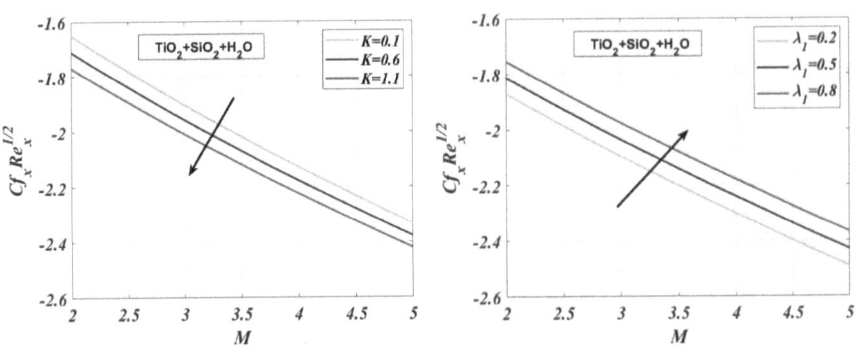

FIGURE 4.12 Variation of skin friction coefficient on M due to change in K and λ_1.

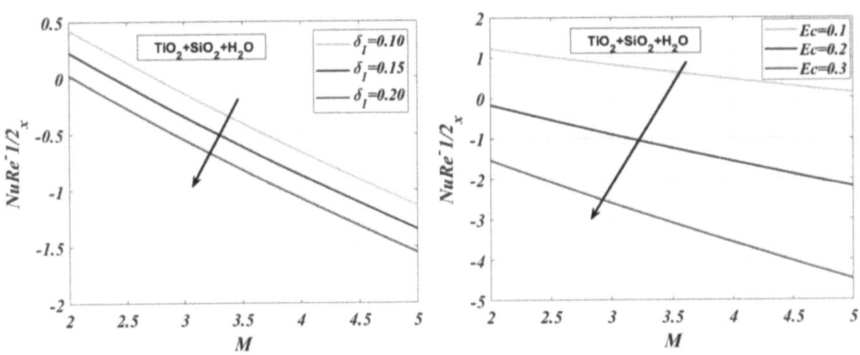

FIGURE 4.13 Variation of heat transfer coefficient on M due to change in δ_1 and Ec.

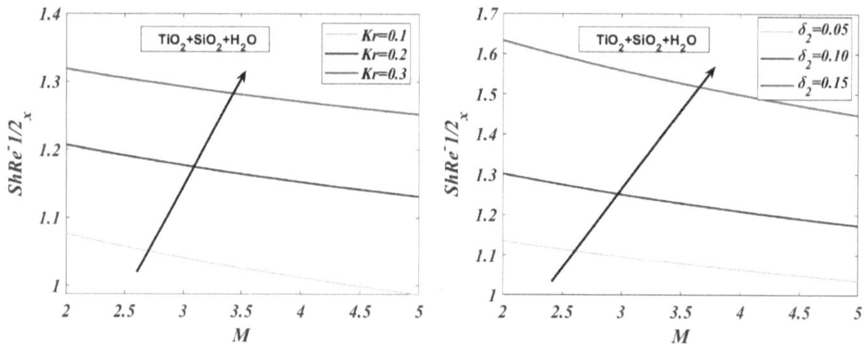

FIGURE 4.14 Variation of mass transfer coefficient on M due to change in Kr and δ_2.

4.5 CONCLUDING REMARKS

This study explores the effect of chemical reaction, buoyancy, and thermal stratification of hybrid nanofluid flow over a vertical stretching plate. The results, obtained using the bvp5c numerical scheme, are presented graphically and reveal key insights into this complex phenomenon.

Among the most relevant findings are that the skin friction coefficient diminishes with higher values of the porous parameter and rises with higher values of the buoyancy parameter. The heat transfer coefficient decreases with increasing values of the thermal stratification parameter and the Eckert number. The mass transfer coefficient enhances with elevated values of both the chemical reaction and solutal stratification parameters. The velocity profile increases with higher values of the buoyancy parameter but decreases with increasing values of the porous and magnetic parameters. The temperature profile increases with higher values of the magnetic and porous parameters but decreases with an increase in the thermal stratification parameter. The concentration profile decreases with increased values of both the solutal stratification and chemical reaction parameters.

Nomenclature

B_o	uniform magnetic field (kg/S^2A)
c	Stretching rate
Ec	Eckert number
Gr	Thermal Grashof parameter
Gc	Solutal Grashof parameter
K_1	rate of reaction constant
K	porosity parameter
M	Magnetic field
Nu	Nusselt number
Pr	Prandtl number
Sc	Schmidt number
Sh	Sherwood number
η	similarity variable

TABLE 4.2
Numerical values of skin friction, Nusselt number, and Sherwood number for different parameters of the hybrid nanofluid

M	K	λ_1	λ_2	Ec	δ_1	Sc	Kr	δ_2	$C_{fx}Re_x^{\frac{1}{2}}$	$NuRe_x^{-\frac{1}{2}}$	$ShRe_x^{-\frac{1}{2}}$
0.2									-1.413600	1.189451	1.248203
0.6									-1.528602	0.640727	1.228125
1.0									-1.636706	0.123220	1.210031
	0.1								-1.318678	0.696543	1.265882
	0.6								-1.502596	0.372918	1.232798
	1.1								-1.668729	0.062100	1.204767
		0.2							-1.636706	0.123220	1.210031
		0.8							-1.467980	0.415526	1.235299
		1.4							-1.305773	0.663965	1.258922
			0.2						-1.636706	0.123220	1.210031
			0.8						-1.448466	0.507196	1.247770
			1.4						-1.266709	0.817671	1.281245
				0.2					-1.649950	1.876756	1.206883
				0.3					-1.645519	1.288436	1.207936
				0.4					-1.641104	0.703937	1.208985
					0.05				-1.633264	0.170966	1.210678
					0.10				-1.636706	0.123220	1.210031
					0.15				0.069742	0.069742	1.209384
						0.3			-1.625946	0.163944	0.690440
						0.6			-1.631038	0.144545	0.926015
						0.9			-1.635393	0.128129	1.141868
							0.1		-1.635422	0.128045	1.147381
							0.2		-1.636706	0.123220	1.210031
							0.3		-1.637887	0.118805	1.269368
								0.05	-1.632698	0.132747	1.184778
								0.10	-1.636706	0.123220	1.210031
								0.15	-1.640714	0.113661	1.238239

α thermal diffusivity
δ_1 thermal stratification parameter
δ_2 solutal stratification parameter
K_r chemical reaction parameter
ρ fluid density (kg/m^3)
ρC_p thermal capacity (J/m^3K)
υ kinematic viscosity (m^2/s)
μ dynamic viscosity $(kg/(ms))$
σ electrical conductivity $\left(\frac{S}{m}\right)$
λ_1 thermal expansion

λ_2 Solutal expansion
f base fluid
nf nanofluid
hnf hybrid nanofluid
ϕ_1, ϕ_2 The volume fraction of nanoparticles

REFERENCES

Amin, A., Munir, S. and Farooq, U., 2023. Investigating flow features and heat/mass transfer in two-layer vertical channel with Gr-TiO2 hybrid nanofluid under MHD and radiation effects. *Journal of Magnetism and Magnetic Materials*, *578*, p.170800.

Anwar, M.I., Ali, M., Lanjwani, H.B. and Shehzad, S.A., 2024. Double stratification impacts on MHD flow of micropolar nanofluid over a stretching surface inserted in porous medium. *International Journal of Modern Physics B*, *38*(26), p.2450359.

Arshad, M. and Hassan, A., 2022. A numerical study on the hybrid nanofluid flow between a permeable rotating system. *The European Physical Journal Plus*, *137*(10), p.1126.

Bashir, A., Hayat, T. and Khan, S.A., 2024. Impact of nonlinear thermal radiation for magnetized dissipative flow of ternary hybrid nanomaterial (Al2 O3-SiO2-Fe3 O4-H2 O). *Journal of Radiation Research and Applied Sciences*, *17*(4), p.101115.

Gul, N., Ramzan, M., Chung, J.D., Kadry, S. and Chu, Y.M., 2020. Impact of hall and ion slip in a thermally stratified nanofluid flow comprising Cu and Al2O3 nanoparticles with nonuniform source/sink. *Scientific Reports*, *10*(1), p.18064.

Hayat, T., Hussain, T., Shehzad, S.A. and Alsaedi, A., 2014. Thermal and concentration stratifications effects in radiative flow of Jeffrey fluid over a stretching sheet. *Plos One*, *9*(10), p.e107858.

Jamaludin, A., Nasir, N.A.A.M., Nazar, R. and Pop, I., 2023. MHD opposing flow of Cu–TiO2 hybrid nanofluid under an exponentially stretching/shrinking surface embedded in porous media with heat source and slip impacts. *Results in Engineering*, *17*, p.101005.

Jamshed, W., Devi, S.U. and Nisar, K.S., 2021. Single phase based study of Ag-Cu/EO Williamson hybrid nanofluid flow over a stretching surface with shape factor. *Physica Scripta*, *96*(6), p.065202.

Junaid, M.S., Aslam, M.N., Khan, M.A., Saleem, S. and Riaz, M.B., 2024. Thermal analysis of a viscoelastic Maxwell hybrid nanofluid with graphene and polythiophene nanoparticles: Insights from an artificial neural network model. *Alexandria Engineering Journal*, *94*, pp.193–211.

Khan, K.A., Vivas-Cortez, M., Ishfaq, K., Javed, M.F., Raza, N., Nisar, K.S. and Abdel-Aty, A.H., 2024. Exploring the numerical simulation of Maxwell nanofluid flow over a stretching sheet with the influence of chemical reactions and thermal radiation. *Results in Physics*, *60*, p.107635.

Khan, K.A., Butt, A.R., Raza, N. and Maqbool, K., 2019. Unsteady magneto-hydrodynamics flow between two orthogonal moving porous plates. *The European Physical Journal Plus*, *134*(1), p.1.

Khan, M.R., Pan, K., Khan, A.U. and Nadeem, S., 2020. Dual solutions for mixed convection flow of SiO2– Al2O3/water hybrid nanofluid near the stagnation point over a curved surface. *Physica A: Statistical Mechanics and its Applications*, *547*, p.123959.

Madiwal, S. and Naduvinamani, N.B., 2024. Heat and mass transformation of Casson hybrid nanofluid (MoS2+ ZnO) based on engine oil over a stretched wall with chemical reaction and thermo-diffusion effect. *Lubricants*, *12*(6), p.221.

Mandal, G. and Pal, D., 2024. Impacts of binary chemical reaction with activation energy on MHD thermally radiating blood based hybrid nanofluid flow through shrinking-stenotic artery: Entropy generation and stability analysis. *Hybrid Advances*, *7*, p.100293.

Masood, S. and Farooq, M., 2021. Influence of thermal stratification and thermal radiation on graphene oxide-Ag/H 2 O hybrid nanofluid. *Journal of Thermal Analysis and Calorimetry*, *143*, pp.1361–1370.

Mishra, A. and Kumar, M., 2019. Viscous dissipation and Joule heating influences past a stretching sheet in a porous medium with thermal radiation saturated by silver–water and copper–water nanofluids. *Special Topics & Reviews in Porous Media: An International Journal*, *10*(2), pp. 171–186.

Mohd Sohut, N.F.H., Soid, S.K., Abu Bakar, S. and Ishak, A., 2022. Unsteady three-dimensional flow in a rotating hybrid nanofluid over a stretching sheet. *Mathematics*, *10*(3), p.348.

Nawaz, M., Nazir, U., Saleem, S. and Alharbi, S.O., 2020. An enhancement of thermal performance of ethylene glycol by nano and hybrid nanoparticles. *Physica A: Statistical Mechanics and Its Applications*, *551*, p.124527.

Rehman, S.U., Mariam, A., Ullah, A., Asjad, M.I., Bajuri, M.Y., Pansera, B.A. and Ahmadian, A., 2021. Numerical computation of buoyancy and radiation effects on MHD micropolar nanofluid flow over a stretching/shrinking sheet with heat source. *Case Studies in Thermal Engineering*, *25*, p.100867.

Shamshuddin, M.D., Shahzad, F., Jamshed, W., Bég, O.A., Eid, M.R. and Bég, T.A., 2023. Thermo-solutal stratification and chemical reaction effects on radiative magnetized nanofluid flow along an exponentially stretching sensor plate: Computational analysis. *Journal of Magnetism and Magnetic Materials*, *565*, p.170286.

Waini, I., Ishak, A. and Pop, I., 2021. Hybrid nanofluid flow towards a stagnation point on an exponentially stretching/shrinking vertical sheet with buoyancy effects. *International Journal of Numerical Methods for Heat & Fluid Flow*, *31*(1), pp.216–235.

5 Nanofluid Flow, Heat, and Mass Transfer at a Stretching Disk

5.1 FLOW PAST A STRETCHING DISK

As just mentioned, laminar boundary layer flow due to stretching disk is of substantial interest for its growing engineering applications and significant implications for those processes. Substantial attention has been paid to analyzing flow and heat transport characteristics of a nanofluid past a stretching disk, as discussed in Abbas et al. (2019), for example. Shateyi and Makinde (2013) investigated flow past a radially stretching disk. They reported the combined effect of viscous dissipation and Joule heating on flow and heat transfer characteristics. Das et al. (2016) investigated entropy generation in a nanofluid flow past a radially stretching disk. They considered a single-phase fluid model for the nanofluid and reported the combined effects of the porous medium and convective boundary conditions on flow and heat transfer. Soid et al. (2017) explored magnetic effects on flow past a radially stretching disk. They found that the skin friction coefficient increases with enhanced values of the magnetic parameter. Kalyani et al. (2017) inspected laminar steady 2D nanofluid flow due to a convectively heated radially stretched disk near a stagnation point in the presence of viscous dissipation and an externally applied magnetic field. They showed that an increased magnetic field caused the temperature to increase but the nanoparticle volume fraction to be reduced. Farooq et al. (2018) explored the combined effects of transpiration and viscous dissipation on hybrid nanofluid flow past a nonlinear radially stretched disk. They also quantified the entropy generation for that flow. By applying an implicit finite difference scheme known as the Keller Box method, they obtained numerical solutions to the problem.

Using spectral relaxation methods, Gangadhar et al. (2018) discussed axisymmetric slip flow of a MHD Powell-Eyring fluid across a radially stretched disk in a porous medium with viscous dissipation. Naganthran et al. (2019) examined the effects of heat generation/absorption on non-Newtonian flow due to a permeable stretching/shrinking disk. They incorporated a Jeffrey fluid model to represent the non-Newtonian fluid behavior. Naqvi et al. (2019) discussed the combined effects of a magnetic field and variable heat generation/absorption on nanofluid flow due to a stretched disk; they used a two-phase model for the nanofluid. They reported that increases in the thermophoresis parameter led to an increase of the nanofluid

DOI: 10.1201/9781003571278-5

temperature while concentration was reduced with higher values of Lewis number, Brownian motion, and suction parameters.

Ahmed et al. (2020) analyzed unsteady non-Newtonian nanofluid flow due to a stretching disk with a Maxwell fluid model near a stagnation point; they showed the consequences of Joule heating on flow and heat transfer. Mahyuddin et al. (2022) quantified heat transfer for Bödewadt flow of a nanofluid past a permeable radially stretched disk. They showed increases to both the local skin friction coefficient and the local Nusselt number with the rise in nanoparticle volume fraction for a radially stretching disk. Khashi'ie et al. (2022) examined the combined effects of viscous dissipation and Joule heating on stagnation point flow of a hybrid nanofluid past a stretching/shrinking disk in the presence of a magnetic field; they showed that the Eckert and Biot numbers significantly influenced the flow and heat transfer. Bilal and Saeed (2022) investigated Sisko hybrid nanofluid flow over a heated permeable shrinking/stretching disk. They noted a large increase in the rate of heat transmission when suction dominated the boundary layer flow.

Here, we shall present results obtained for nanofluid flow, heat, and mass transfer over a radially stretching disk. No one has yet reported on the combined effects of variable stretching and suction for such flows.

5.1.1 Formulation of the Problem

Consider a steady 2D laminar viscous incompressible nanofluid flow past a radially stretching disk. The disk is elongated by the velocity $u_w = ar$ radially, a being a constant (see Figure 5.1).

The disk is placed at $z = 0$ at a uniform temperature, T_w having constant nanoparticle concentration, C_w. The symbols T_∞ and C_∞ stand for free stream temperature and free stream concentration, respectively.

Using standard notation, the governing equations are described as follows (from Soid et al. (2017) and Naqvi et al. (2019)):

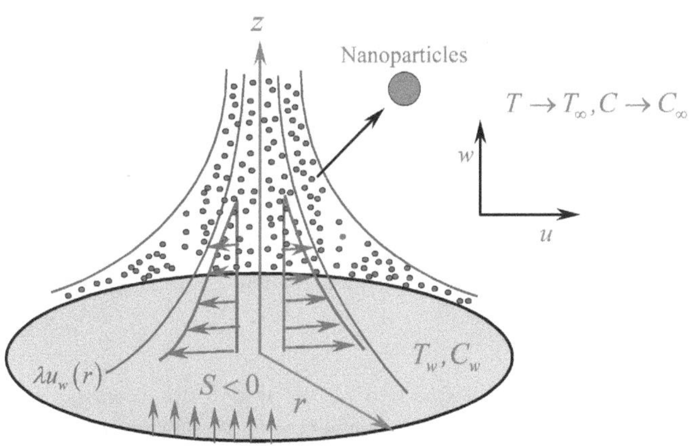

FIGURE 5.1 Sketch of the physical flow problem.

$$\frac{\partial u}{\partial r}+\frac{u}{r}+\frac{\partial w}{\partial z}=0, \tag{5.1}$$

$$u\frac{\partial u}{\partial r}+w\frac{\partial u}{\partial z}=\nu\frac{\partial^2 u}{\partial z^2}, \tag{5.2}$$

$$u\frac{\partial T}{\partial r}+w\frac{\partial T}{\partial z}=\frac{\kappa}{\rho c_p}\frac{\partial^2 T}{\partial z^2}+\tau\left[D_B\frac{\partial T}{\partial z}\frac{\partial C}{\partial z}+\frac{D_T}{T_\infty}\left(\frac{\partial T}{\partial z}\right)^2\right], \tag{5.3}$$

$$u\frac{\partial C}{\partial r}+w\frac{\partial C}{\partial z}=D_B\frac{\partial^2 C}{\partial z^2}+\frac{D_T}{T_\infty}\frac{\partial^2 T}{\partial z^2}, \tag{5.4}$$

where u and w are the velocities of fluid in the r and z directions, respectively. The symbol ρ is the liquid density, kinematic viscosity is ν, T represents the temperature of the liquid, the thermal conductivity is κ, specific heat capacitance is ρc_p, τ is the ratio of the heat capacitance of the nanoparticles to that of the base fluid. The symbol D_B represents the diffusion coefficient due to Brownian motion and D_T is the diffusion coefficient due to thermophoresis. Finally, C is the nanoparticle concentration. The corresponding situations at the boundary for this problem are described in Soid et al. (2017) and Naqvi et al. (2019) as:

$$u=\lambda u_w(r)=\lambda ar,\ w=w_0,\ T=T_w,\ C=C_w \text{ at } z=0, \tag{5.5}$$

$$u\to 0,\ T\to T_\infty,\ C\to C_\infty \text{ as } z\to\infty, \tag{5.6}$$

where $\lambda>0$ is the stretching parameter, $w_0<0$ represents suction, and $w_0>0$ represents blowing.

5.1.2 Similarity Transformations

In order to obtain the solution of this problem, the partial differential equations (5.2)–(5.4) are transformed into ordinary differential equations by using an appropriate similarity alteration as follows

$$\eta=\sqrt{\frac{a}{\nu}}z, u=ar\,f'(\eta), w=-2\sqrt{a\nu}\,f(\eta),\ \theta=\frac{T-T_\infty}{T_w-T_\infty}, \phi=\frac{C-C_\infty}{C_w-C_\infty}. \tag{5.7}$$

Equation (5.1) is automatically satisfied by the transformations in Eq. (5.7). Next, inserting Eq. (5.7) into Eqs. (5.2)–(5.4), we get the following equations, with Table 5.1 summarizing the dimensionless parameters that emerge in the analysis, along with their definitions.

$$f''' + ff'' - f'^2 = 0, \tag{5.8}$$

$$\frac{1}{Pr}\theta'' + 2f\theta' + \left[Nb\ \theta'\phi' + Nt\,\theta'^2\right] = 0, \tag{5.9}$$

$$\phi'' + 2Le\,f\phi' + \frac{Nt}{Nb}\theta'' = 0. \tag{5.10}$$

Again, inserting (5.7) into (5.5)–(5.6), we have,

$$f'(0) = \lambda, f(0) = S,\ \theta(0) = 1,\ \phi(0) = 1, \tag{5.11}$$

$$f'(\eta) \to 0, \theta(\eta) \to 0, \phi(\eta) \to 0 \text{ as } \eta \to \infty. \tag{5.12}$$

To simply the expressions, multiple dimensionless numbers are defined as shown here:

$Nb = \tau D_B (C_w\text{-}C_\infty)/ v,$
$Nt = D_T (T_w - T_\infty)/ T_\infty v,$
$Le = \text{v}/\text{D}_\text{B},$
$Pr = v\rho c_p/\kappa,$
$S = -w_0 / 2\sqrt{(av)}.$

5.1.3 Numerical Solutions and Discussion of the Results

To get a clear insight into the physical problem, numerical computations have been carried out for various values of different parameters involved in this problem. To verify the accuracy of the numerical scheme, the results obtained in this study are compared with the available results of Makinde et al. (2016) and Soid et al. (2017) and found to be in excellent agreement with the results. The comparison is presented in Table 5.1.

Figures 5.2a–c present velocity, temperature, and concentration for the variations of the stretching parameter λ. Initially, the fluid velocity rises with increasing values of the stretching parameter λ in the presence of blowing, but later, the fluid velocity diminishes with the rising values of λ. This behavior is displayed in Figure 5.2a. The wall shear stress $f''(0)$ diminishes with the rise in the stretching parameter λ, as shown in Table 5.2. The temperature also diminishes with the rising values of the stretching

TABLE 5.1
Validation of $f''(0)$ results for $\lambda = 1$ and $S = 0$

Makinde et al., (2016), $M = 0$	Soid et al., (2017), $M = 0$	Present results
-1.17372	-1.1737	-1.17372

TABLE 5.2
Values for $f''(0)$, $-\theta'(0)$, and $-\varphi'(0)$ for various parameter values

λ	S	Pr	Nb	Nt	Le	$f''(0)$	$-\theta'(0)$	$-\phi'(0)$
0.3	-0.1	5.2	0.1	0.1	0.5	-0.163888	0.578585	-0.182151
0.5						-0.364434	0.878949	-0.335823
0.8						-0.758048	1.243030	-0.541408
	-0.2					-0.683318	0.789203	-0.144109
	-0.3					-0.616876	0.453024	0.123276
		6.2				-	0.417621	0.164462
		8.2				-	0.344956	0.244290
			0.2			-	0.295539	0.296140
			0.3			-	0.251445	0.311636
				0.2		-	0.220282	0.330691
				0.3		-	0.193188	0.361404
					0.8	-	0.146131	0.544337
					1.0	-	0.127053	0.621868

parameter λ in the presence of blowing, as shown in Figure 5.2b. The temperature gradient at the wall $\theta'(0)$ reduces with the rise in the stretching parameter λ, as evident from Table 5.2. Though the concentration initially rises with increasing stretching parameter, later, the concentration decreases with the further increases in λ in the presence of blowing. This behavior is displayed in Figure 5.2(c). A concentration overshoot is noted near the surface. The concentration gradient at the surface $\varphi'(0)$ rises with λ, as evident from Table 5.2.

Figure 5.3a shows that the fluid velocity diminishes with the rising values of suction/blowing parameter S. For increasing blowing wall shear stress, $f''(0)$ rises, as shown in Table 5.2. The temperature also decreases with rising values of suction/blowing parameter S. Temperature and blowing strength are positively correlated, as evident from Figure 5.3b. The temperature gradient at the wall $\theta'(0)$ also rises for higher values of blowing (Table 5.2). Fluid concentration decreases with a rise in suction/blowing parameter S (see Figure 5.3c). Near the surface, concentration overshoot is noted. The concentration gradient at the surface $\varphi'(0)$ diminishes with rising strength of blowing (see Table 5.2).

Figure 5.4a illustrates the effects of Brownian motion parameter Nb on the thermal field in the presence of blowing (S = -0.2). This figure shows the variation of temperature profiles for increasing values of the Brownian motion parameter (Nb) in the presence of blowing and when the other parameters are kept constant. Clearly, as Nb increases, the temperature also increases. This behavior is related to the decreasing thickness of the boundary layer. Brownian motion arises due to the random movement of free nanoparticles suspended in the liquid. Owing to the random motion of the

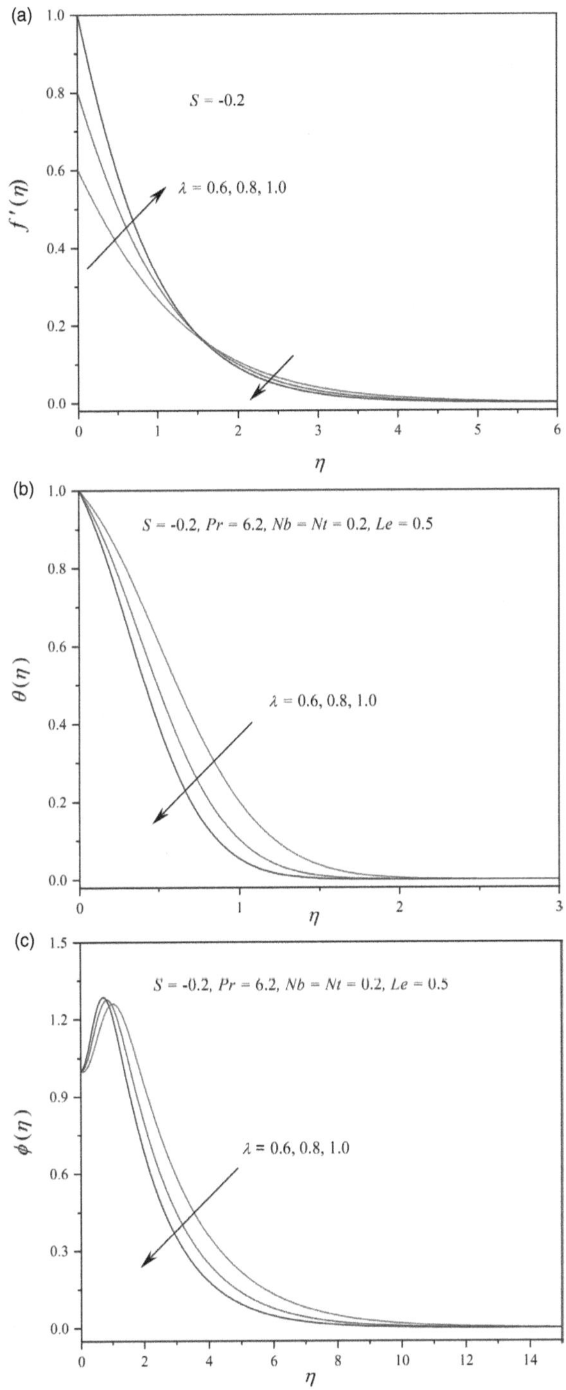

FIGURE 5.2 (a) Velocity, (b) temperature, and (c) concentration profiles for variable values of the stretching parameter λ.

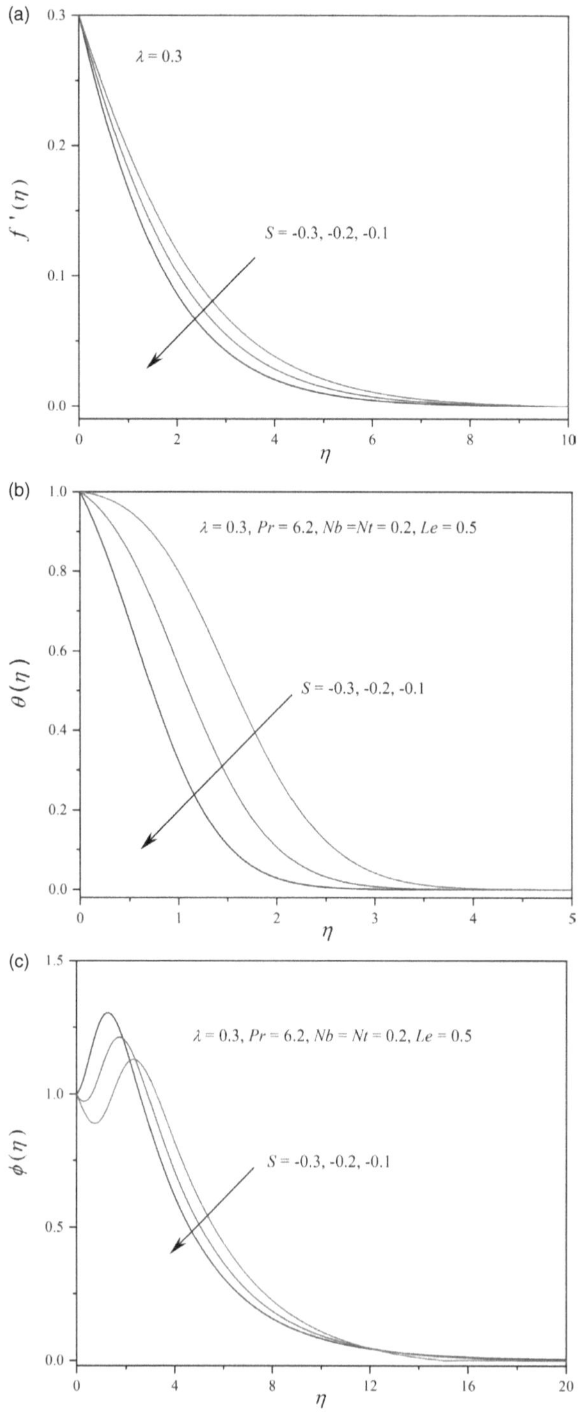

FIGURE 5.3 (a) Velocity, (b) temperature, and (c) concentration profiles for various values of suction/blowing parameter S.

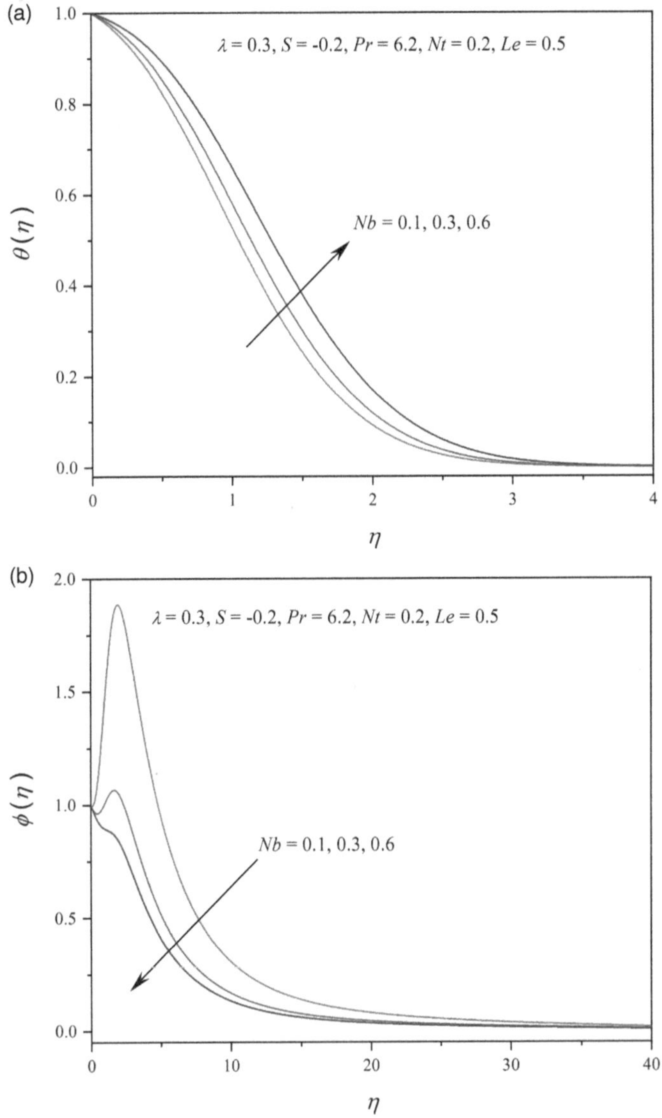

FIGURE 5.4 (a) Temperature and (b) concentration profiles for variable values of the Brownian motion parameter *Nb*.

particles, the kinetic energy increases. Accordingly, the temperature of the nanoliquid is enhanced. The heat transfer rate at the surface $\theta'(0)$ rises with the rise in *Nb* (see Table 5.2). The concentration profiles for several increasing values of *Nb* in the presence of blowing are exhibited in Figure 5.4b. From this figure, it is clear that concentration reduces with the rise in *Nb*.

Concentration overshoot is noted near the surface of the disk for $Nb = 0.1$ and 0.3, but no overshoot is noted for higher values of *Nb*. The mass transfer rate at the surface

$\varphi\,'(0)$ diminishes for enhanced values of *Nb* (see Table 5.2). The fluid temperature and the concentration rise for rising values of thermophoresis parameter *Nt* (Figure 5.5a and 5.5b). For growing values of *Nt*, the rate of heat transfer at the surface $\theta\,'(0)$ increases (see Table 5.2). Concentration overshoot is noted near the surface of the disk. Higher overshoot values of the concentration profile are noted for higher values of *Nt* (Figure 5.5b). The mass transfer rate at the surface $\varphi\,'(0)$ diminishes with a rise in *Nt* (see Table 5.2).

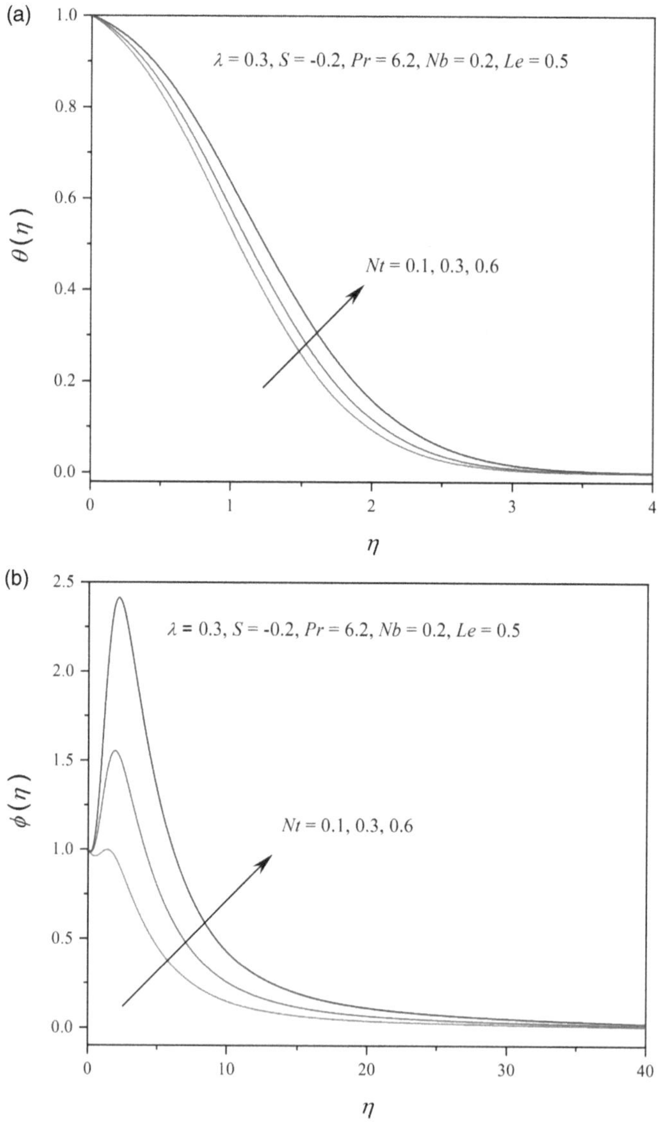

FIGURE 5.5 (a) Temperature and (b) concentration profiles for various values of thermophoresis parameter *Nt*.

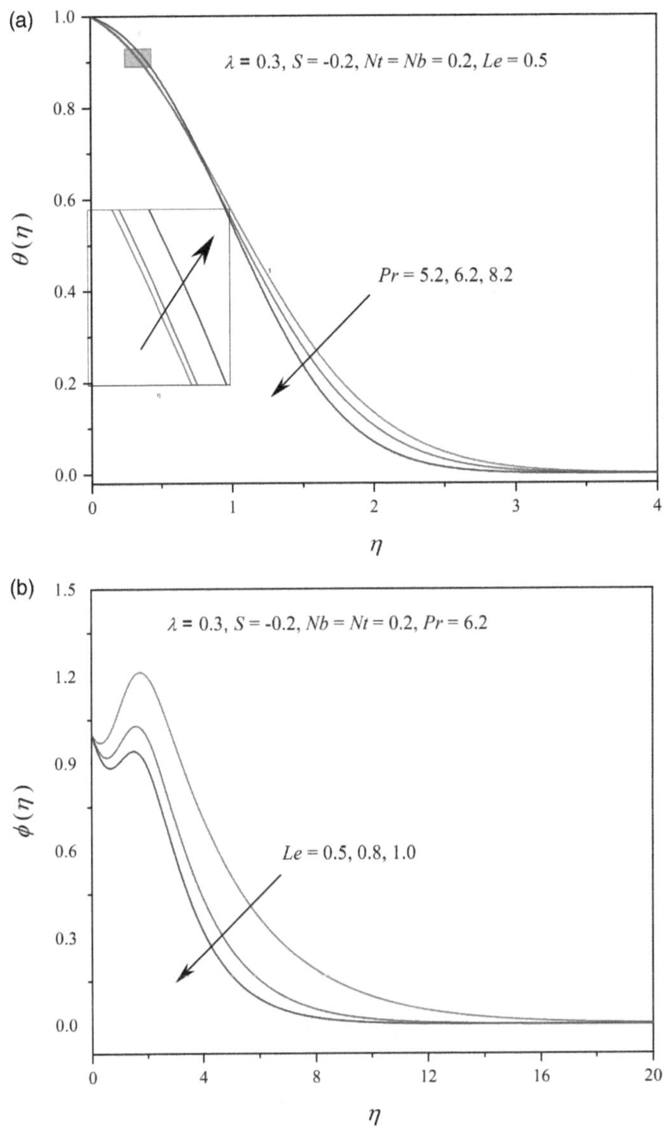

FIGURE 5.6 (a) Temperature profiles for various values of Prandtl number *Pr* and (b) concentration profiles for multiple values of Lewis number *Le*.

Though the temperature profile is initially enhanced with growing values of the Prandtl number *Pr*, the profile later reverses course and decreases with further rises in *Pr* (see Figure 5.6a). The wall temperature gradient $\theta'(0)$ increases with rising *Pr*, but the concentration gradient at the surface $\varphi'(0)$ diminishes (see Table 5.2). A reduction in concentration is also noted for a rise in the Lewis number, *Le* (Figure 5.6b). Near the surface of the disk, concentration overshoot occurs. The rate of heat transfer at the surface, $\theta'(0)$, rises while the mass transfer rate decreases with rising values of *Le*.

5.2 FLOW PAST A ROTATING STRETCHING DISK

As already discussed, boundary-layer flow due to a rotating disk is a classical fluid dynamics problem. It has been considered by many researchers. Some practical uses include viscometers, rotating bodies, rotors, centrifugal pumps, turbines, spinning disks, and fans, as discussed in Bilal et al. (2020). Flows due to a rotating disk have widespread applications in physics, engineering, automobiles, industries, and other fields of science (Bilal et al., 2022). Boundary layer behavior over a rotating stretching disk in a free stream has additional practical applications such as thermal power generating systems, rotating machinery fields, computer storage devices, heat and mass exchangers, geothermal industry, gas turbine rotators, chemical processes, medical machinery, electronic devices, and air cleaning systems (Tulu, 2023).

MHD flows due to rotating disks are of both theoretical and practical interest. With this motivation, Khan et al. (2017) inspected the entropy generation for Powell-Eyring fluid flow due to a radially stretching rotating disk in the presence of a magnetic field. They also reported the heat dissipation, Soret, Dufour and Joule heating effects on flow, heat, and mass transfer. Khan et al. (2017) inspected heat and mass transfer for MHD Maxwell fluid flow past a stretching rotating disk. With a Runge-Kutta-Felberg fourth- and fifth-order (RKF45) integration scheme using the MAPLE software, they obtained the numerical solutions of the system.

Khan et al. (2018) reported the combined effects of chemical reaction and radiation for flow past a rotating disk. Using the Chebyshev collocation technique, they solved the Navier-Stokes, Maxwell, and energy equations that governed the physical situation. Gowthami et al. (2020) developed a mathematical model to represent the flow over a rotating stretching disk embedded in an orthotropic porous medium. They employed a shooting technique using a 4th-order Runge-Kutta method and Newton's method to obtain the numerical solutions. Ramzan et al. (2021) reported on the impacts of an oscillating magnetic field and hall current on flow past a rotating stretching disk; they considered nano-ferrofluid in their investigation. Reddy et al. (2021) examined the MHD flow of a hybrid nanofluid past a stretching rotating disk; they reported the effects of Arrhenius energy on flow and heat transfer. Hussain et al. (2022) presented the effects of thermal radiation on nanofluid flow due to a rotating stretching disk. They also explored the consequences of generalized slip at the boundary.

Here, we expand upon these previous research efforts by presenting results of nanofluid flow past a permeable stretching rotating disk with convective boundary condition.

5.2.1 Mathematical Modeling

Let us consider a non-rotating frame to study the three-dimensional axisymmetric flow of incompressible nanofluid over a rotating stretching disk. The disk is rotating with angular velocity Ω about the vertical axis z; the rotating disk is aligned along $z = 0$. Let the nanofluid fill the half-space $z > 0$. The disk is considered to be stretchable in the radial direction with a rate $s_1 r$ where s_1 is a constant. The physical structure is shown in Figure 5.7.

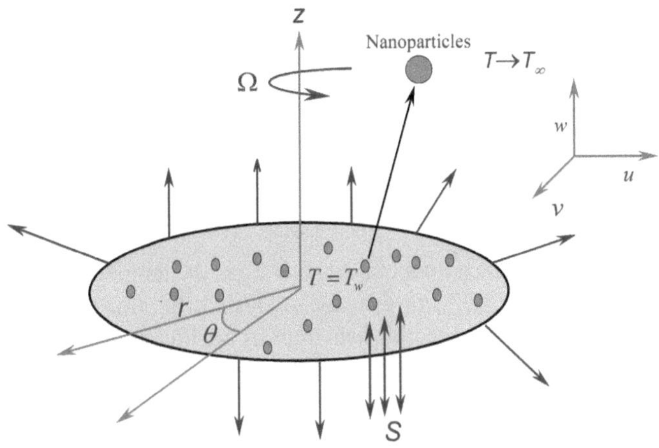

FIGURE 5.7 Sketch of the physical flow problem.

Employing Oberbeck-Boussinesq approximations, mass, momentum, and thermal energy conservation equations in the boundary layer are:

$$\frac{\partial u}{\partial r}+\frac{u}{r}+\frac{\partial w}{\partial z}=0, \tag{5.13}$$

$$\rho_{nf}\left(u\frac{\partial u}{\partial r}+w\frac{\partial u}{\partial z}-\frac{\upsilon^2}{r}\right)=-\frac{\partial p}{\partial r}+\mu_{nf}\frac{\partial^2 u}{\partial z^2}, \tag{5.14}$$

$$\rho_{nf}\left(u\frac{\partial \upsilon}{\partial r}+w\frac{\partial \upsilon}{\partial z}+\frac{u\upsilon}{r}\right)=\mu_{nf}\frac{\partial^2 \upsilon}{\partial z^2}, \tag{5.15}$$

$$\rho_{nf}\left(u\frac{\partial w}{\partial r}+w\frac{\partial w}{\partial z}\right)=-\frac{\partial p}{\partial z}+\mu_{nf}\left[\frac{1}{r}\frac{\partial}{\partial r}\left(r\frac{\partial w}{\partial r}\right)+\frac{\partial^2 w}{\partial z^2}\right], \tag{5.16}$$

$$\left(\rho c_p\right)_{nf}\left(u\frac{\partial T}{\partial r}+w\frac{\partial T}{\partial z}\right)=\kappa_{nf}\frac{\partial^2 T}{\partial z^2}, \tag{5.17}$$

where u, υ, and w are the components of velocity respectively in r, θ, and z directions, p is the pressure, T is the temperature, $v_{nf}=\mu_{nf}/\rho_{nf}$ is the kinematic viscosity, μ_{nf} is the viscosity, ρ_{nf} is the nanofluid density, κ is the thermal conductivity, c_p represents the specific heat at constant pressure, and $(\rho c_p)_{nf}$ is the specific heat capacitance of the nanofluid.

The appropriate boundary conditions for the problem are

$$u = U_w(r) = s_1 r,\ \upsilon = \Omega r,\ w = w_c, -\kappa_{bf}\frac{\partial T}{\partial z} = -h_c(T - T_w)\ at\ z = 0, \tag{5.18a}$$

$$u \to 0, \upsilon \to 0, T \to T_\infty \text{ as } z \to \infty, \tag{5.18b}$$

Here, T_w is the uniform surface temperature, and T_∞ is the temperature far away from the surface. We assume $T_w > T_\infty$ however the analysis if the flow of heat is reversed. The symbol h_c stands for the convective heat transfer coefficient, while w_c indicates the strength of suction/blowing. Here, subscript labels *nf*, *bf*, and *sp* stand for nanofluid, base fluid, and nanoparticles, respectively.

We also have the following relationships for the thermophysical properties of the nanofluid:

$$\frac{\rho_{nf}}{\rho_{bf}} = (1-\phi) + \phi\frac{\rho_{sp}}{\rho_{bf}}, \tag{5.19a}$$

$$\frac{(\rho c_p)_{nf}}{(\rho c_p)_{bf}} = (1-\phi) + \phi\frac{(\rho c_p)_{sp}}{(\rho c_p)_{bf}}, \tag{5.19b}$$

$$\mu_{nf} = \frac{\mu_{bf}}{(1-\phi)^{\frac{5}{2}}}, \tag{5.19c}$$

$$\frac{\kappa_{nf}}{\kappa_{bf}} = \frac{\frac{\kappa_{sp}}{\kappa_{bf}} + 2 - 2\phi\left(1 - \frac{\kappa_{sp}}{\kappa_{bf}}\right)}{\frac{\kappa_{sp}}{\kappa_{bf}} + 2 + \phi\left(1 - \frac{\kappa_{sp}}{\kappa_{bf}}\right)}. \tag{5.19d}$$

5.2.2 Solutions Procedure

Let us now introduce the following dimensionless variables:

$$\eta = z\sqrt{\frac{s_1}{\nu_{bf}}},$$

$$u = s_1 r f(\eta),$$

$$v = s_1 r g(\eta),$$

$$w = \sqrt{s_1 \nu_{bf}}\, h(\eta),$$

$$\frac{p}{p_\infty} = -s_1 \nu_{bf} P(\eta),$$

$$\theta = \frac{T - T_\infty}{T_w - T_\infty}. \tag{5.20}$$

Inserting Eq. (5.20) into Eqs. (5.14) – (5.17) yields:

$$2f(\eta) + h'(\eta) = 0, \tag{5.21}$$

$$\frac{\nu_{nf}}{\nu_{bf}} f''(\eta) - h(\eta) f'(\eta) - f^2(\eta) + g^2(\eta) = 0, \tag{5.22}$$

$$\frac{\nu_{nf}}{\nu_{bf}} g''(\eta) - h(\eta) g'(\eta) - 2f(\eta) g(\eta) = 0, \tag{5.23}$$

$$\frac{\nu_{nf}}{\nu_{bf}} h''(\eta) + \frac{\rho_{bf}}{\rho_{nf}} P'(\eta) - h(\eta) h'(\eta) = 0, \tag{5.24}$$

$$\frac{\dfrac{\kappa_{nf}}{\kappa_{bf}}}{\dfrac{(\rho c_p)_{nf}}{(\rho c_p)_{bf}}} \frac{1}{Pr} \theta''(\eta) - \theta'(\eta) h(\eta) = 0. \tag{5.25}$$

The boundary conditions take the following form:

$$f(0) = 1,\ g(0) = C,\ h(0) = S,\ \theta'(0) = \frac{\kappa_{bf}}{\kappa_{nf}} Bi\left[\theta(0) - 1\right], \tag{5.26}$$

$$f(\eta)\to 0,\ g(\eta)\to 0,\ \theta(\eta)\to 0,\ P(\eta)\to 0\ as\ \eta\to\infty. \tag{5.27}$$

To simply the expressions, multiple dimensionless numbers are defined as shown here:

$$Pr = \nu_{bf}\,(\rho c_p)_{bf}/\kappa_{bf},$$

$$C = \Omega / \sigma_1,$$

$$Bi = h_c / \kappa_{\beta\phi}\sqrt{(\nu_{\beta\phi}/s_1)},$$

$$S = \omega_\chi/\sqrt{(\sigma_1\nu_{bf})},$$

where C represents the rotational parameter.

Equations (5.21) and (5.25), along with boundary conditions (5.26)–(5.27) are solved numerically.

5.2.3 Discussion of the Numerical Results

As previously discussed, numerical computations have been carried out using bvp4c in MATLAB®, which can account for various values of the parameters of this problem.

To validate the simulation, our results are compared with the available results of Fang (2007) and presented in Table 5.3, which shows a perfect agreement.

Additional numerical results are presented in Figs. 5.8(a)–5.11(b) and in Tables 5.5–5.7.

The effects of the rotational parameter C on radial velocity $f(\eta)$ are exhibited in Figure 5.8(a). From this figure, it is observed that with an increasing rotational parameter, the radial velocity initially rises but later reverses direction. One can conclude that disk rotation has a significant effect, as noted near the disk surface. Next, we consider the impact of the rotational parameter C on the tangential velocity $g(\eta)$. The tangential velocity of the fluid rises with the rise in rotational parameter (see Figure 5.8b). This effect is not significant near the edge of the boundary layer. A constant tangential velocity is noted for $C = 0$, as shown in Figure 5.8b; $C = 0$ implies there is no rotation. The axial velocity $h(\eta)$ diminishes with increasing values of the rotation parameter C (Figure 5.8c). With the rotational motion, more fluids are

TABLE 5.3
Results of $f'(0)$ and $h''(0)$ when $\varphi = 0$, $C = 0$, and $S = 0$ along with Fang (2007)

	Fang (2007)	Present Study
$f'(0)$	-1.173721	-1.737215
$h''(0)$	2.347442	2.347443

TABLE 5.4
Skin friction coefficients and Nusselt numbers for viscous fluids and for Ag-water nanofluid for several values of C when $S = 0.6$, $Bi = 0.5$, and $Pr = 6.2$

	$\varphi = 0$		Ag -water nanofluid with $\varphi = 0.03$	
C	$C_f Re_r^{1/2}$	$Nu_r Re_r^{-1/2}$	$C_f Re_r^{1/2}$	$Nu_r Re_r^{-1/2}$
0	0.7908	0.2940	0.9660	0.3039
0.5	1.0490	0.2958	1.2850	0.3058
1.0	1.9406	0.3006	2.3918	0.3108
1.5	3.7609	0.3071	4.6627	0.3176
2.0	6.8919	0.3144	8.5848	0.3250
2.5	11.7503	0.3217	14.6894	0.3324
3.0	18.3289	0.3289	23.5271	0.3395
3.5	28.3839	0.3357	35.6586	0.3462
4.0	41.0419	0.3420	51.6505	0.3524

directed towards the disk surface, so the axial velocity rises in the negative z direction. With higher rotational speeds, the centrifugal forces are enhanced. The temperature also diminishes in this case (see Figure 5.8d). Larger values of C means the angular velocity is higher than the stretching velocity.

With increasing nanoparticle volume fraction ϕ, the radial velocity and tangential velocity both diminish, but the axial velocity rises (see, respectively, Figures 5.9a, 5.9b, and 5.9c). Initially, near the surface, the fluid temperature diminishes, but away from the disk, the surface temperature rises with the enhanced values of nanoparticle volume fraction φ (see Figure 5.9d). The rate of heat transport is enhanced with increased nanoparticle volume fraction.

Figure 5.10a presents the influences of the suction parameter S on radial velocity. It is observed that the radial velocity, $f(\eta)$, of the fluid is significantly enhanced for rising values of S. The tangential velocity, $g(\eta)$ (see Figure 5.10b) and axial velocity, $h(\eta)$ (see Figure 5.10c). Both rise with the growing values of S. In addition, the boundary layer thickness rises with the suction parameter S. The dimensionless temperature $\theta(\eta)$ is also enhanced (Figure 5.10d).

The Biot number, Bi, indicates the relative ease of heat transfer by convection and conduction. Values of the Biot number determine whether a transient heating or cooling process can be approximated by a lumped temperature parameter. Figure 5.11a shows that with the rise in Biot number Bi, temperature also rises. However, the temperature diminishes with increasing values of Prandtl number Pr (see Figure 5.11b) as the thermal diffusivity diminishes. Initially, near the disk, a slight increase in the temperature is noted as evident in Figure 5.11b.

The skin friction coefficient rises significantly with the rise in rotational parameter C (see Table 5.4). However, higher increases in the skin friction coefficient are noted

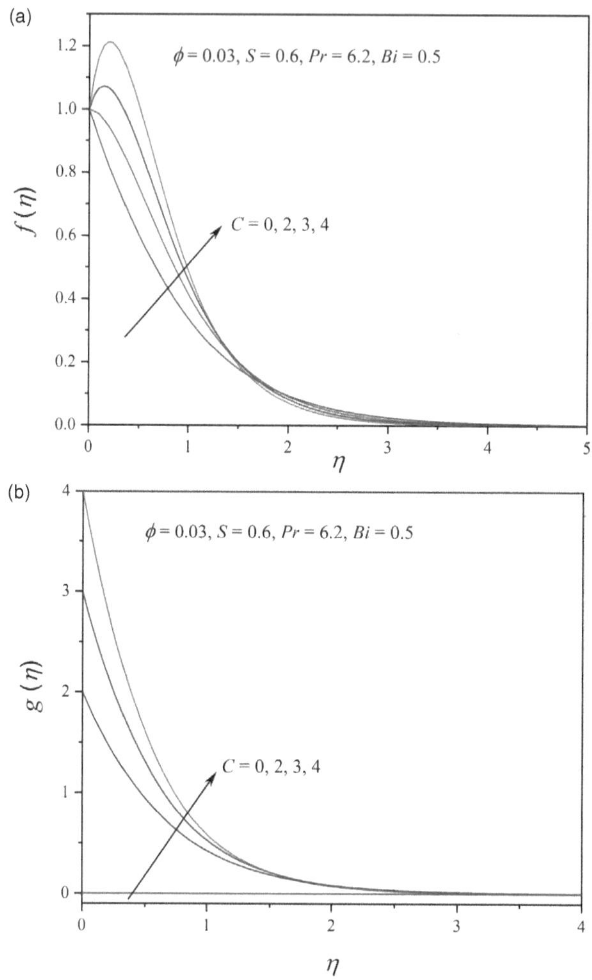

FIGURE 5.8 Effects of rotational parameter C on (a) radial velocity, (b) tangential velocity, (c) axial velocity, and (d) temperature profiles.

TABLE 5.5
Nusselt number for viscous fluid and Ag-water nanofluid for several values of *Bi* when $C = 2.0$, $S = 0.6$, and $Pr = 6.2$

Bi	Base Fluid, $\varphi = 0$	Ag -water nanofluid with $\varphi = 0.03$
0.2	0.16179	0.16456
0.4	0.27168	0.27957
0.6	0.35118	0.36448
0.8	0.41138	0.42974
1.0	0.45853	0.48147
1.2	0.49648	0.52348

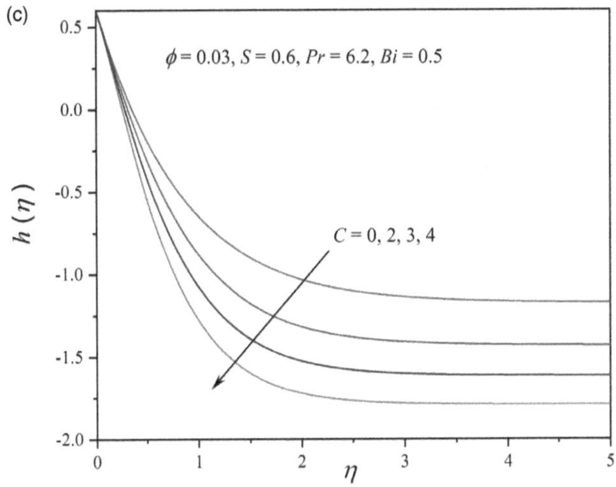

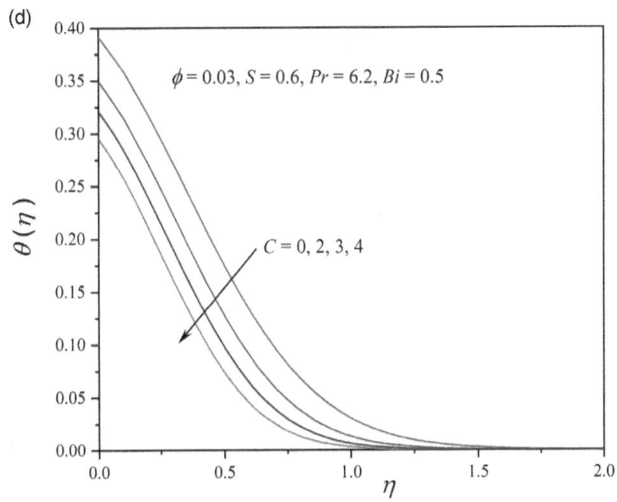

FIGURE 5.8 (Continued)

TABLE 5.6
Skin friction coefficient and Nusselt number of Ag-water nanofluid for several values of φ when $S = 0.6$, $C = 2.0$, $Bi = 0.5$, and $Pr = 6.2$

ϕ	$C_f Re_r^{1/2}$	$Nu_r Re_r^{-1/2}$
0.00	6.81864	0.31438
0.01	7.46118	0.31803
0.03	8.58480	0.32499
0.05	9.69509	0.33155
0.07	10.79864	0.33772
0.10	12.45309	0.34632

for the Ag-water nanofluid compared to a viscous fluid (Table 5.4). The lowest skin friction coefficient is noted for $C = 0$. The Nusselt number rises slowly with the rise in rotational parameter C for both cases of a viscous fluid and the Ag-water nanofluid (see Table 5.4). For $C = 0$, the lowest heat transfer coefficient is noted.

The heat transfer coefficient rises slowly with the Biot number for both viscous fluids and Ag-water nanofluids (see Table 5.5). Higher values of Nusselt number are observed for Ag-water nanofluid compared to that for a viscous fluid (Table 5.5).

The skin friction coefficient also rises rapidly with the rise in nanoparticle volume fraction ϕ but increases in the heat transfer coefficient are not so significant (Table 5.6). The lowest values of the skin friction and heat transfer coefficients are observed for ordinary viscous fluid, i.e., for $\varphi = 0$. These results are consistent with our earlier findings.

5.3 FLOW PAST AN UNSTEADY STRETCHING DISK

Until now, in this chapter, the flow has been steady. However, flow and heat transfer may be unsteady due to impulsive stretching of a surface or by a step change in the temperature of the surface, for example. Unsteadiness potentially impacts flow characteristics and heat transport. Several researchers investigated unsteady flow and heat transfer passing over differently shaped geometrical bodies. Considering the decelerating case, Fang and Hua (2012) examined the unsteady flow of a viscous fluid over a stretchable rotating disk. Rashidi et al. (2013) discussed entropy generation MHD for flow due to unsteady stretching rotating disk. Hobiny et al. (2015) obtained numerical solutions for unsteady flow due to a nonlinearly stretching and rotating disk. Turkyilmazoglu (2018) examined the flow and heat transfer due to an unsteady moving rotating vertical disk. Waini et al. (2019) examined the unsteady flow of a hybrid nanofluid passing over a stretching/shrinking disk. They also presented a stability analysis to identify stable solutions.

Prasad et al. (2020) reported the consequences of an external magnetic field and suction/blowing on flow past an unsteady stretching rotating disk. They also reported the combined influences of temperature-dependent fluid properties and convective boundary conditions. Bhandari et al. (2020) investigated the unsteady flow features of ferrofluid between two stretchable rotating disks. Ahmed et al. (2020) analyzed the effects of Joule heating on the unsteady flow of a non-Newtonian nanofluid past a stretching disk near a stagnation point; they considered a Maxwell fluid to represent the non-Newtonian flow behavior. Sarkar and Shaoo (2021) obtained dual solutions for unsteady flow due to a stretching and rotating disk in the presence of a magnetic field. They presented a temporal stability analysis to point out the stable solution. Punith Gowda et al. (2021) examined particle deposition in an unsteady flow due to a rotating disk moving vertically. Khan et al. (2021) analyzed the unsteady flow of a hybrid nanofluid due to radially stretching/shrinking permeable disk. They considered Cu-Al_2O_3-water in their investigation. Mandal and Shit (2021) discussed entropy generation for the unsteady flow of a bi-viscous nanofluid due to a stretching and rotating disk; they reported the characteristics of thermal radiation on convective heat transfer.

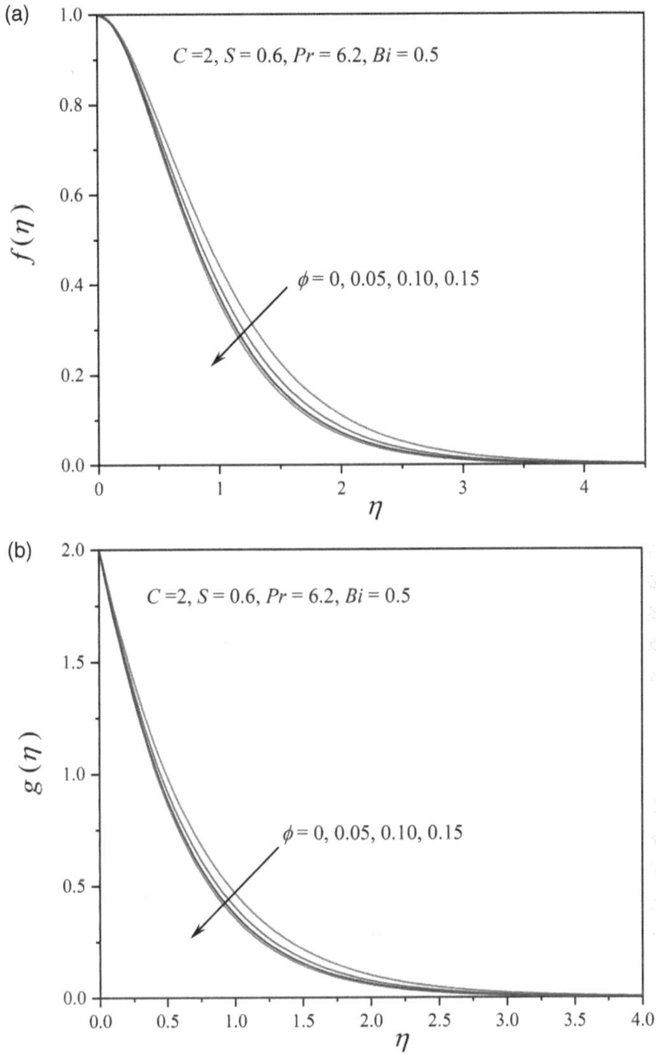

FIGURE 5.9 Effects of nanoparticle volume fraction φ on (a) radial velocity, (b) tangential velocity, (c) axial velocity, and (d) temperature profiles.

Rashad et al. (2021) inspected bioconvective unsteady non-Newtonian nanofluid flow due to a stretching disk near a stagnation point. They considered an Oldroyed-B fluid model to represent non-Newtonian fluid behavior and discussed the combined impacts of activation energy and temperature-based thermal conductivity on flow and heat transfer. Recently, using a ternary hybrid Casson nanofluid, Khan et al. (2021) examined the unsteady flow due to a nonlinear stretching disk in a porous medium and reported the impacts of slip and externally applied magnetic field.

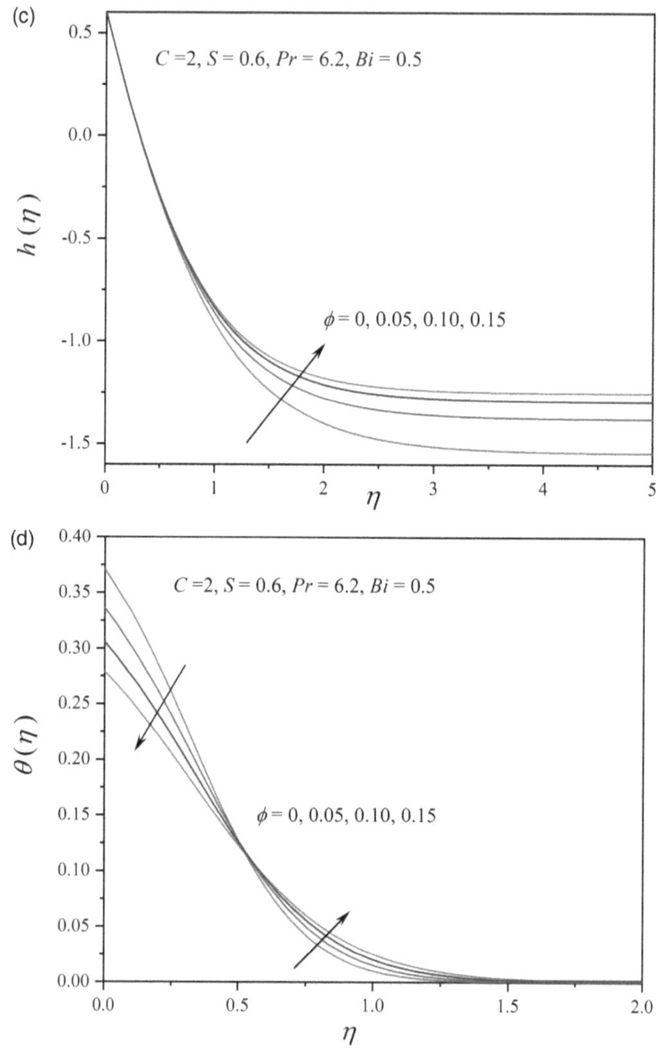

FIGURE 5.9 (Continued)

Fluid flow and heat transfer near a stagnation point deserve special attention. Not many research works are available on unsteady flow due to a stretching disk near a stagnation point. So, we are interested in presenting the flow and heat transfer characteristics of a hybrid nanofluid flow under those conditions.

5.3.1 Mathematical Modeling

Let us consider flow past a stretching disk. Here, the surface is located at $z = 0$, and the flow occurs in the upper halfplane (see Figure 5.12). The governing equations for the

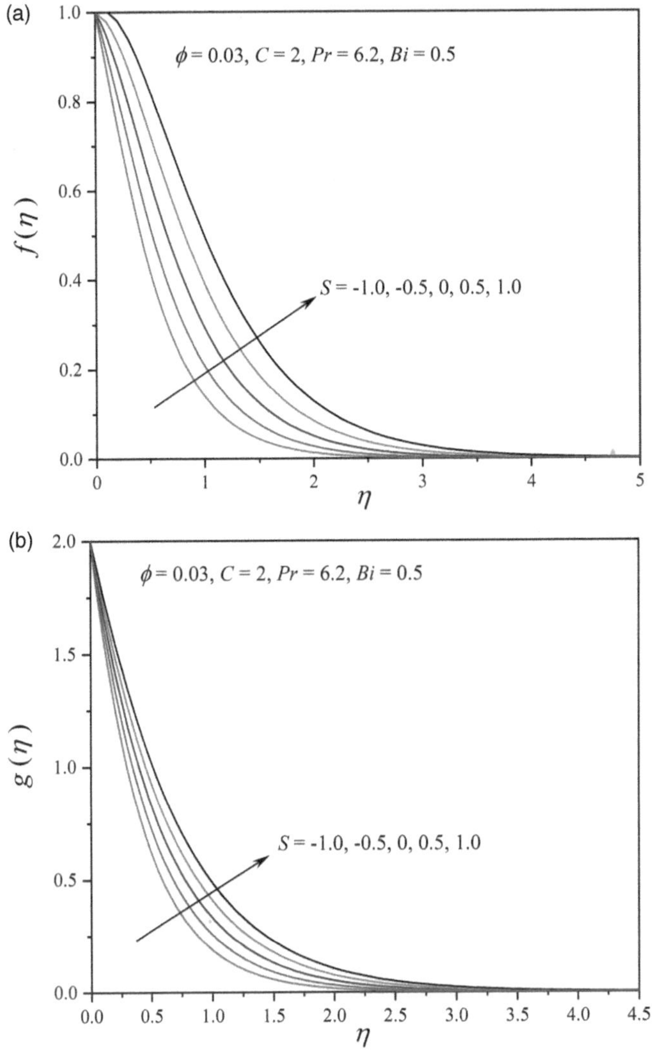

FIGURE 5.10 Effects of suction parameter S on (a) radial velocity, (b) tangential velocity, (c) axial velocity, and (d) temperature profiles.

unsteady Agrawal flow of a hybrid nanofluid are given by Waini et al. (2021), Rosca and Pop (2017), and Weidman (2014) as:

$$\frac{1}{r}\frac{\partial}{\partial r}(ru)+\frac{\partial w}{\partial z}=0, \tag{5.28}$$

$$\frac{\partial u}{\partial t}+u\frac{\partial u}{\partial r}+w\frac{\partial u}{\partial z}=-\frac{1}{\rho_{hnf}}\frac{\partial p}{\partial r}+\frac{\mu_{hnf}}{\rho_{hnf}}\left(\frac{\partial^2 u}{\partial r^2}+\frac{1}{r}\frac{\partial u}{\partial r}+\frac{\partial^2 u}{\partial z^2}-\frac{u}{r^2}\right), \tag{5.29}$$

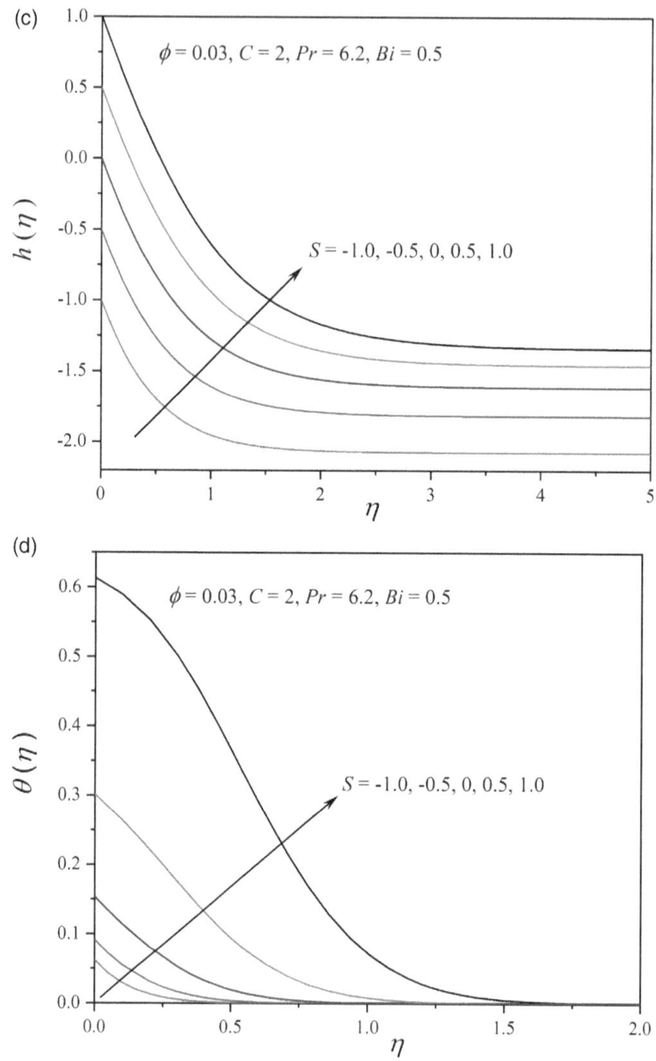

FIGURE 5.10 (Continued)

$$\frac{\partial w}{\partial t}+u\frac{\partial w}{\partial r}+w\frac{\partial w}{\partial z}=-\frac{1}{\rho_{hnf}}\frac{\partial p}{\partial z}+\frac{\mu_{hnf}}{\rho_{hnf}}\left(\frac{\partial^2 w}{\partial r^2}+\frac{1}{r}\frac{\partial w}{\partial r}+\frac{\partial^2 w}{\partial z^2}\right), \tag{5.30}$$

$$\frac{\partial T}{\partial t}+u\frac{\partial T}{\partial r}+w\frac{\partial T}{\partial z}=\frac{\kappa_{hnf}}{\left(\rho c_p\right)_{hnf}}\left(\frac{\partial^2 T}{\partial r^2}+\frac{1}{r}\frac{\partial T}{\partial r}+\frac{\partial^2 T}{\partial z^2}\right). \tag{5.31}$$

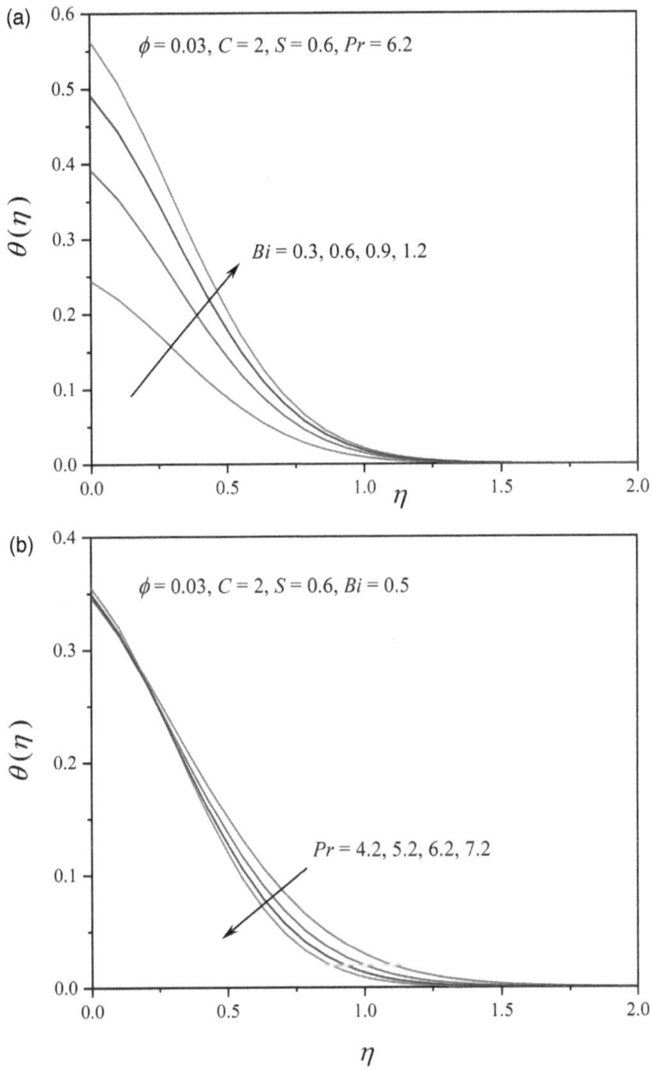

FIGURE 5.11 Effects of (a) Biot number, *Bi*, and (b) Prandtl number, *Pr*, on temperature profiles.

Here, $u_w\,(r,t)$ and $u_e\,(r,t)$ represent respectively the surface and free stream velocities. T_w and T_∞ are the uniform surface temperature and free stream temperature, respectively. The symbol p is the pressure, ρ_{hnf} and μ_{hnf} are the densities and viscosities of the hybrid nanofluid, k_{hnf}, $(\rho c_p)_{hnf}$ are the thermal conductivity and specific heat capacitance of the hybrid nanoliquid, respectively. In the r and z directions, the velocity components are u and w, respectively, and T is the temperature of the hybrid nanofluid.

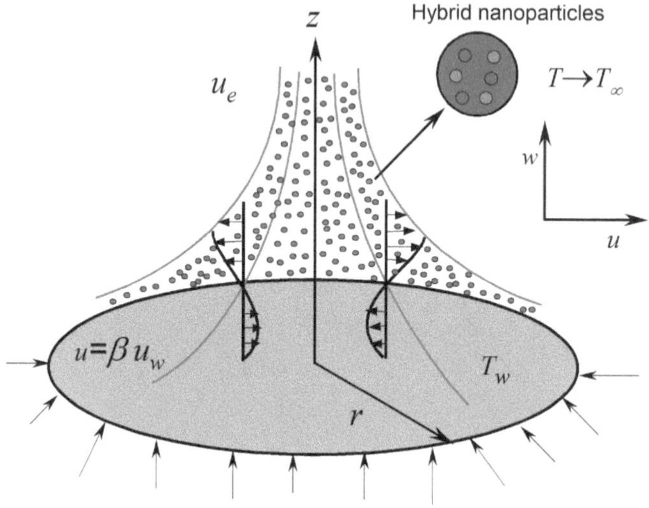

FIGURE 5.12 Sketch of the physical flow problem.

The boundary conditions for this type of flow are given by Waini et al. (2021) as:

$$u = u_w(r,t) = \beta U_w(r,t), w = 0, T = T_w \ at \ z = 0, \tag{5.32}$$

$$\frac{\partial u(r,t)}{\partial z} \to \frac{\partial u_e(r,t,z)}{\partial z}, T \to T_\infty \ as \ z \to \infty, \tag{5.33}$$

where $U_w(r,t) = \dfrac{a^{\frac{2}{3}} f^{\frac{1}{3}} r}{(1-\alpha t)}$ and $u_e(r,t,z) = \dfrac{2arz}{(1-\alpha t)^{\frac{3}{2}}}$.

Here, a is a constant having the dimension of length^{-1} time^{-1}. It measures the strength of Agrawal flow. The term $\beta = 0$ corresponds to a static surface while $\beta > 0$ represents the stretching disk. α is a positive constant with dimensions of time^{-1}.

The thermophysical properties of the hybrid nanofluid and its correlation with nanofluid are given in Tables 5.7 and 5.8. Here, *hnf* and *f* signify, respectively, the hybrid nanofluid and the base fluid, φ_1 and φ_2, represent the Al_2O_3 and Cu nanoparticle concentrations, respectively.

5.3.2 Similarity Transformations

Eliminating pressure p from Equations (5.29) and (5.30) allows the following equation to be obtained

$$\frac{\partial^2 u}{\partial z \partial t} + \frac{\partial u}{\partial z}\frac{\partial u}{\partial r} + w\frac{\partial^2 u}{\partial z \partial r} + \frac{\partial w}{\partial z}\frac{\partial u}{\partial z} = \frac{\mu_{hnf}}{\rho_{hnf}}\left(\frac{1}{r}\frac{\partial^2 u}{\partial z \partial r} + \frac{\partial^3 u}{\partial z^3} - \frac{1}{r^2}\frac{\partial u}{\partial z}\right). \tag{5.34}$$

TABLE 5.7
Correlations for nanofluids and hybrid nanofluids

Properties	Relations
Density (ρ)	$\rho_{hnf} = \left(1-\phi_{hnf}\right)\rho_f + \phi_1\rho_{s1}$
Heat Capacity (ρc_p)	$\left(\rho C_p\right)_{hnf} = \left(1-\phi_{hnf}\right)\left(\rho C_p\right)_f + \phi_1\left(\rho C_p\right)_{s1} + \phi_2\left(\rho C_p\right)_{s2}$
Dynamic viscosity (μ)	$\frac{\mu_{hnf}}{\mu_f} = \frac{1}{\left(1-\phi_{hnf}\right)^{2.5}}$
Thermal conductivity (k)	$\frac{k_{hnf}}{k_f} = \left[\frac{\left(\frac{\phi_1k_1+\phi_2k_2}{\phi_{hnf}}\right)+2k_f+2\left(\phi_1k_1+\phi_2k_2\right)-2\phi_{hnf}k_f}{\left(\frac{\phi_1k_1+\phi_2k_2}{\phi_{hnf}}\right)+2k_f-\left(\phi_1k_1+\phi_2k_2\right)+\phi_{hnf}k_f}\right]$

Sources: Takabi and Salehi (2014) and Oztop and Abu-Nada (2008).

TABLE 5.8
Thermophysical properties of Al_2O_3, Cu, and H_2O

Thermophysical Properties	H_2O (*f*)	Nanoparticles	
		$Al_2O_3\left(\phi_1\right)$	Cu $\left(\phi_1\right)$
C_p	4179	765	385
k	0.613	40	400
ρ	997.1	3970	8933

Source: Oztop and Abu-Nada (2008).

Let us consider the following similarity transformations.

$$u = \frac{a^{\frac{2}{3}}\nu_f^{\frac{1}{3}}r}{(1-\alpha t)}f'(\eta),$$

$$w = -\frac{2a^{\frac{2}{3}}\nu_f^{\frac{2}{3}}}{\sqrt{(1-\alpha t)}}f(\eta),$$

$$\eta=\left(\frac{a}{\nu_f}\right)^{\frac{1}{3}}\frac{z}{\sqrt{(1-\alpha t)}},$$

$$\theta=\frac{T-T_\infty}{T_w-T_\infty}. \tag{5.35}$$

Using the similarity transformations given by Eq. (5.35), we obtain the transformed forms of the Eqs. (5.31) and (5.34) as follows:

$$\frac{\frac{\mu_{hnf}}{\mu_f}}{\frac{\rho_{hnf}}{\rho_f}}f'''+2ff''-f'^2-A\left(\frac{\eta}{2}f''+\frac{3}{2}f'\right)=0, \tag{5.36}$$

$$\frac{1}{Pr}\frac{\left(\rho c_p\right)_f}{\left(\rho c_p\right)_{hnf}}\left[\frac{k_{hnf}}{k_f}+\frac{4}{3}R\right]\theta''+2f\theta'-A\frac{\eta}{2}\theta'=0. \tag{5.37}$$

With the help of Eq. (5.35), the boundary conditions of Eqs. (5.32)–(5.33) now become

$$f'(0)=\beta, f(0)=0, \theta(0)=0, \tag{5.38}$$

$$f''(\eta)\to 2, \theta(\eta)\to 0 \quad \text{as } \eta\to\infty. \tag{5.39}$$

Subject to the boundary conditions (5.38)–(5.39), Equations (5.36)–(5.37) are numerically solved using Mathematica software.

5.3.3 Results and Discussion

For the verification of the numerical results, the values of the skin friction coefficient for a steady flow of an ordinary viscous fluid are compared with the results of Waini et al. (2021) and found to be in excellent agreement. The validation results are listed in Table 5.9.

To elucidate the physical behavior, variations of fluid velocity, velocity gradient, and temperature are exhibited in Figures 5.13a–5.16. Additionally, skin friction coefficients and heat transfer coefficients for various values of different parameters and are presented in Tables 5.10–5.12. Dual solutions are obtained for a stretching disk ($\beta > 0$), while unique solutions exist for $\beta = 0$ (a static disk). Fluid velocity

TABLE 5.9
Values of $Re_r^{1/2} C_f$ for different values of β with $A = 0$ and $\varphi_{hnf} = 0\%$

	Waini et al. (2021)		Present Study	
β	1st Solution	2nd Solution	1st Solution	2nd Solution
1	1.2614	-1.3296	1.2614	-1.3296
0.8	1.5039	-0.9668	1.5039	-0.9668
0.6	1.7054	-0.6437	1.7054	-0.6437
0.4	1.8607	-0.3664	1.8607	-0.3664
0.2	1.9625	-0.1450	1.9626	-0.1450
0.0	2.0000	-	2.0000	-

TABLE 5.10
Skin friction coefficient and Nusselt numbers for several values of φ_1 when $\varphi_2\phi_2 = 0.01$, $A = 0.02$, $\beta = 1.2$, and $Pr = 6.2$

	$C_f Re_r^{1/2}$		$Nu_r Re_r^{-1/2}$	
φ_1	1st Solution	2nd Solution	1st Solution	2nd Solution
0.0	0.19339	-2.73720	3.65998	3.17609
0.01	0.19636	-2.81009	3.71011	3.21153
0.05	0.22039	-3.11551	3.91135	3.33245
0.1	0.28265	-3.53363	4.68870	3.46668
0.12	0.31970	-3.71416	4.27473	3.51975
0.15	0.39096	-4.00138	4.43735	3.59052

TABLE 5.11
Skin friction coefficients and Nusselt numbers for several values of A when $\varphi_1 = 0.01$, $\varphi_2 = 0.01$, $\beta = 1.2$, and $Pr = 6.2$

	$C_f Re_r^{1/2}$		$Nu_r Re_r^{-1/2}$	
A	1st Solution	2nd Solution	1st Solution	2nd Solution
0.0	0.28474	-2.82716	3.72573	3.20459
0.01	0.24050	-2.81873	3.71793	3.20763
0.02	0.19636	-2.81009	3.71011	3.21153
0.03	0.15234	-2.80116	3.70227	3.21251
0.04	0.10845	-2.79180	3.69439	3.21585
0.05	0.06468	-2.78178	3.68649	3.22581
0.06	0.02104	-2.77071	3.67857	3.22527
0.07	-0.02139	-2.75769	3.67062	3.22515

TABLE 5.12
Skin friction coefficient (C_f) and Nusselt number (Nu_r) for several values of β when $\varphi_1 = 0.01$, $\varphi_2 = 0.01$, $A = 0.02$, and $Pr = 6.2$

	$C_f Re_r^{1/2}$		$Nu_r Re_r^{-1/2}$	
β	1st Solution	2nd Solution	1st Solution	2nd Solution
0.8	1.47848	-1.02822	2.91051	0.15432
1.0	1.21507	-1.41930	3.12813	1.24820
1.2	0.91142	-1.75766	3.33265	2.45804
1.5	0.38759	-2.55771	3.61919	3.08875
1.8	-0.21074	-3.33741	3.88580	3.42284
2.0	-0.64773	-3.89338	4.05427	3.61571

diminishes with the rising values of A for both solutions, as shown in Figure 5.13a. The velocity gradient also diminishes for both solutions with rising values of A (see Figure 5.13b). The temperature rises slightly for the first solution. In contrast, for the second solution, the temperature near the disk surface diminishes with the rise in unsteady parameter A but rises in the second solution (Figure 5.13c).

In the first branch of the solution, the fluid velocity diminishes with the rise in φ_{hnf}, but for the second branch, the fluid velocity rises. This behavior is displayed in Figure 5.14a. The velocity gradient also diminishes for the first solution with the rise in φ_{hnf} but this term rises in the second solution under similar conditions. A rise in temperature is noted for the first solution for rising values of φ_{hnf}. However, the temperature rises near the surface of the disk and later diminishes with the rising values of φ_{hnf} for the second solution (Figure 5.14c).

For rising values of the stretching parameter β, the fluid velocity rises for both solutions (Figure 5.15a). The velocity gradient diminishes with the rise in β for the first solution. For the second solution, the velocity gradient initially diminishes but later increases with a rise in the stretching parameter β (Figure 5.15b). Fluid temperature is found to reduce with the rising values of the stretching parameter β for both branches of solutions (Figure 5.15c).

Near the surface of the disk, fluid temperature increases; however, away from the surface, temperature diminishes for the first solution with a rise in the Prandtl number. The temperature also diminishes with the rising values of the Prandtl number for the second branch of the solution, as displayed in Figure 5.16.

With the rising values of nanoparticle volume fraction φ_1 the skin friction coefficient increases slightly for the first solution, but it diminishes for the second solution, as listed in Table 5.10. The heat transfer coefficient rises rapidly for both solutions with φ_1. The lowest skin friction and heat transfer coefficients are observed for $\varphi_1 = 0$ (Table 5.10). The skin friction coefficient diminishes rapidly in the first solution as the unsteady parameter A (Table 5.11), but the opposite trend is noted in the second solution. The heat transfer coefficient diminishes slightly for the first solution with the rise in unsteady parameter A but rises slowly in the second solution. The maximum skin friction and heat transfer coefficient occur for steady flow in the first solution. It

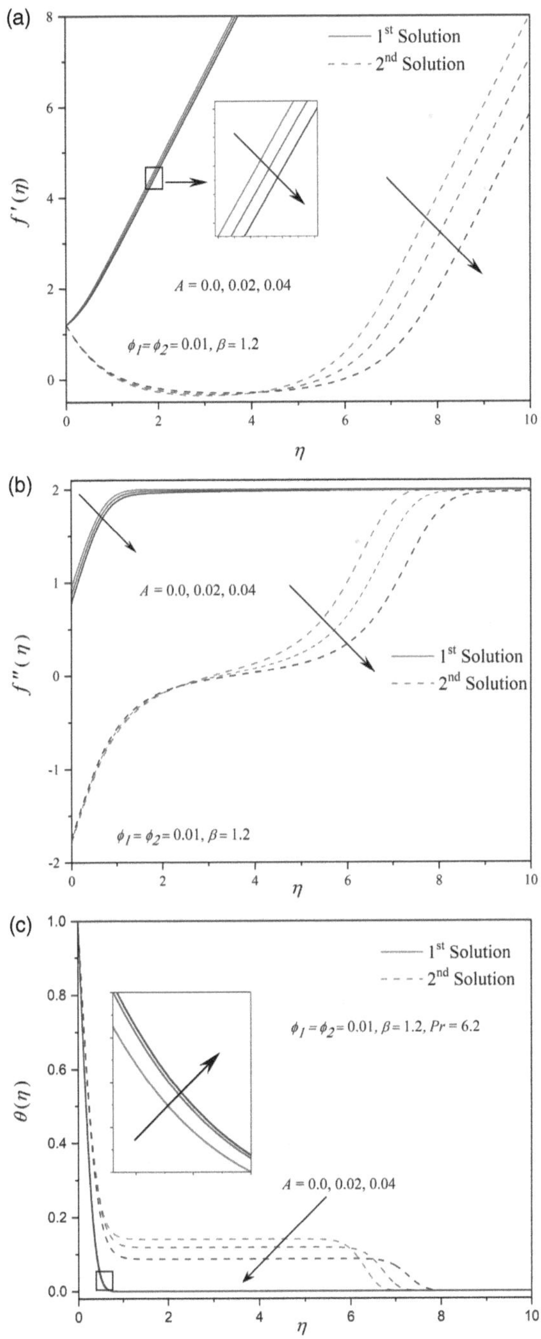

FIGURE 5.13 (a) Velocity, (b) velocity gradient, and (c) temperature profiles for several values of unsteadiness parameter A.

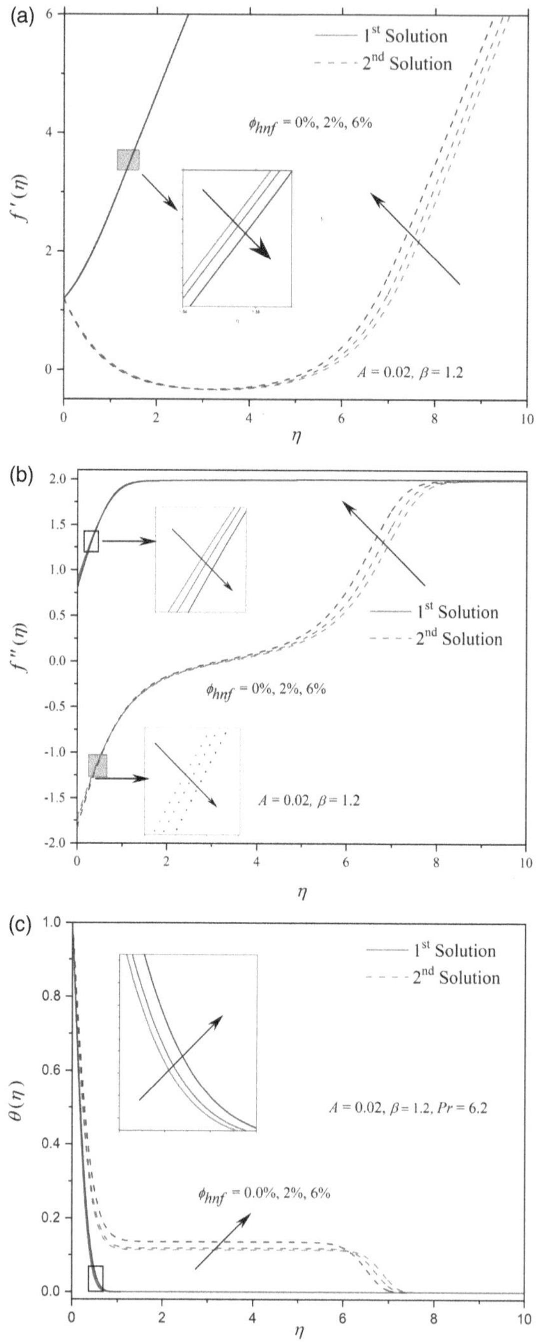

FIGURE 5.14 (a) Velocity, (b) velocity gradient, (c) temperature profiles for several values of different values of φ_{hnf}.

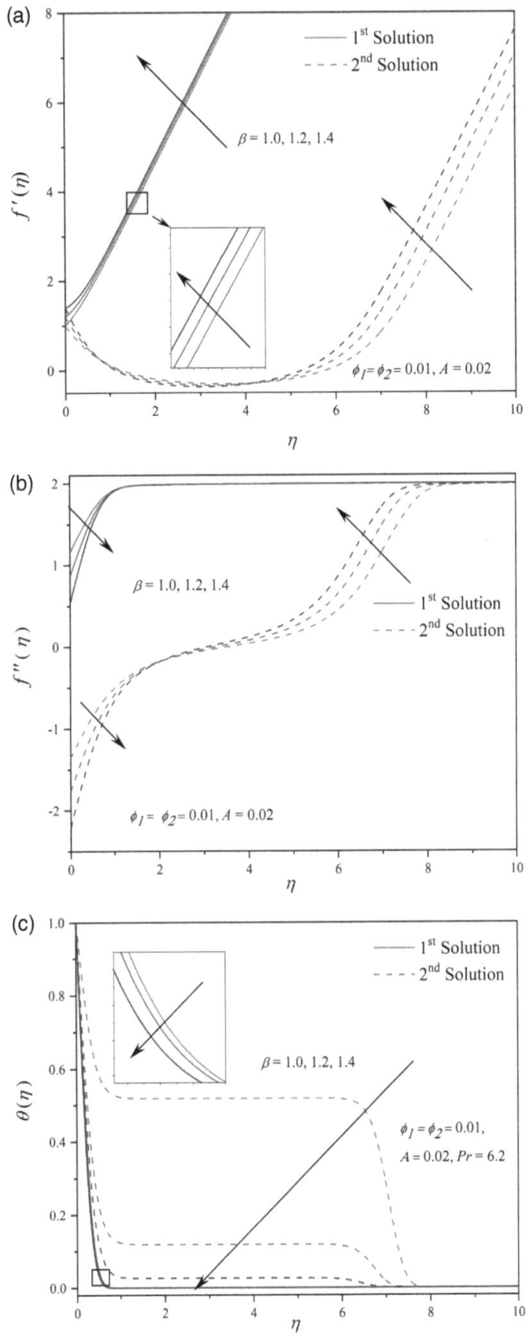

FIGURE 5.15 (a) Velocity, (b) velocity gradient, and (c) temperature profiles for several values of β.

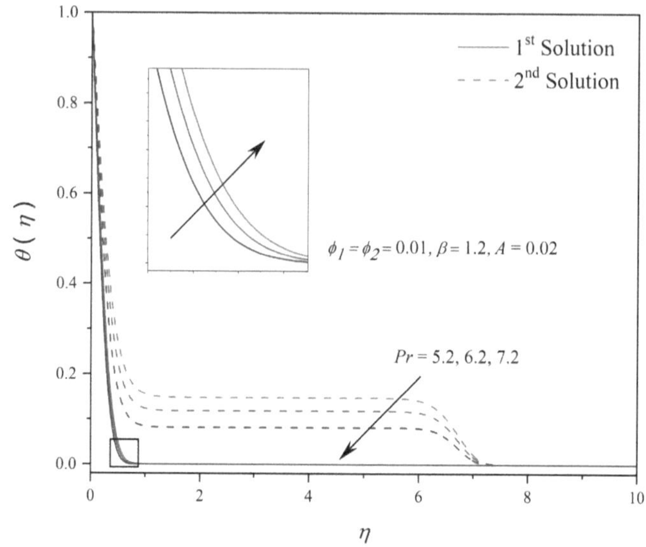

FIGURE 5.16 Temperature profiles for the variations of Prandtl number *Pr*.

is to be noted that the first solution is stable, whereas the second solution is unstable. From a physical point of view, the first solution is most important; however, from a theoretical perspective, both solutions have value.

The skin friction coefficient initially increases with the rise in the stretching parameter up to $\beta = 1$, but later, it diminishes with rising values of the stretching parameter β for the first branch solution (Table 5.12). But for the second solution, the skin friction coefficient diminishes for all values of β . The heat transfer coefficient rises with an increase in the stretching parameter β for the first solution and diminishes in the second solution (see Table 5.12).

5.4 CONCLUDING REMARKS

Nanofluid flow, heat, and transfer due to a stretching disk have been discussed in this chapter. In the first section, nanofluid flow due to a radially stretching disk has been discussed in detail including the effects of Brownian motion and thermophoresis. In the next section, nanofluid flow due to stretching and rotating is investigated in detail. In the last section, hybrid nanofluid flow due to an unsteady stretching disk is analyzed. It was found that fluid flow, heat transfer, and mass transfer are significantly affected by the parameters involved in our study. The results obtained in our study are relevant to the broader applications in industrial sectors and warrant future studies of the problem under different physical situations.

Nomenclature

A	unsteadiness parameter
Bi	Biot number

C nanoparticle concentration
c_p specific heat (J/kg K)
C_∞ free stream concentration
C_w surface concentration
D_B Brownian diffusion coefficient
D_T thermophoresis diffusion coefficient
Le Lewis number
Nb Brownian motion parameter
Nt thermophoresis parameter
Pr Prandtl number
R thermal radiation parameter
S suction/blowing parameter
T temperature of the liquid (K)
T_∞ free stream temperature (K)
T_w surface temperature (K)
u, v velocity component in x, y direction (m/s)
φ non-dimensional concentration
η similarity variable
κ liquid thermal conductivity ($Wm^{-1}K^{-1}$)
ν kinematic viscosity (m^2/s)
θ non-dimensional fluid temperature
ρ liquid density ($kg\ m^{-3}$)
τ ratio of heat capacitance of the nanoparticles to that of the base fluid

REFERENCES

Abbas, Z., Jafar, M.A. and Hasnain, J., 2019. Asymptotic analysis of MHD viscous fluid flow due to a rotating disc and a radially stretching-shrinking disc with Navier slip condition. *Physica Scripta*, *95*(2), p.025204.

Ahmed, A., Khan, M., Ahmed, J., Hafeez, A. and Iqbal, Z., 2020. Unsteady stagnation point flow of maxwell nanofluid over stretching disk with joule heating. *Arabian Journal for Science and Engineering*, *45*, pp.5529–5540.

Bhandari, A. and Husain, A., 2020. Unsteady flow and heat transfer of a ferrofluid between two rotating, vertical moving, and stretching disks. *Heat Transfer*, *50*(2), pp.1570–1589.

Bilal, M. and Saeed, A., 2022. Numerical computation for the dual solution of Sisko hybrid nanofluid flow through a heated shrinking/stretching porous disk. *International Journal of Ambient Energy*, *43*(1), pp.8802–8811.

Bilal, M., Khan Marwat, D.N. and Ali, A., 2020. The flow of a viscous fluid over an infinite rotating and porous disk with stretching (shrinking) effects. *Revista mexicana de física*, *66*(2), pp.171–179.

Bilal, M., Marwat, D.N.K. and Alam, A., 2022. Diffusion of heat and mass in flow of viscous fluid over a moving porous and rotating disk. *Advances in Mechanical Engineering*, *14*(8), p.16878132221119917.

Das, S., Chakraborty, S., Jana, R.N. and Makinde, O.D., 2016. Entropy analysis of nanofluid flow over a convectively heated radially stretching disk embedded in a porous medium. *Journal of Nanofluids*, *5*(1), pp.48–58.

Fang, T., 2007. Flow over a stretchable disk. *Physics of Fluids*, *19*(12), p.128105.

Fang, T. and Tao, H., 2012. Unsteady viscous flow over a rotating stretchable disk with deceleration. *Communications in Nonlinear Science and Numerical Simulation*, *17*(12), pp.5064–5072.

Farooq, U., Afridi, M.I., Qasim, M. and Lu, D., 2018. Transpiration and viscous dissipation effects on entropy generation in hybrid nanofluid flow over a nonlinear radially stretching disk. *Entropy*, *20*(9), p.668.

Gangadhar, K., Pasumarti, R.S.B. and Makinde, O.D., 2018, November. Spectral relaxation method for Powell-Eyring fluid flow past a radially stretching heated disk surface in a porous medium. In *Defect and Diffusion Forum* (Vol. 387, pp. 575–586). Trans Tech Publications Ltd.

Gowda, R.P., Kumar, R.N., Aldalbahi, A., Issakhov, A., Prasannakumara, B.C., Rahimi-Gorji, M. and Rahaman, M., 2021. Thermophoretic particle deposition in time-dependent flow of hybrid nanofluid over rotating and vertically upward/downward moving disk. *Surfaces and Interfaces*, *22*, p.100864.

Gowthami, K., Prasad, P.H., Mallikarjuna, B. and Makinde, O.D., 2020. Hydrodynamic flow between rotating stretchable disks in an orthotropic porous medium. *Songklanakarin Journal of Science and Technology*, *42*, pp.391–397.

Hobiny, A., Ahmad, B., Alsaedi, A., Jalil, M. and Asghar, S., 2015. Similarity solution for flow over an unsteady nonlinearly stretching rotating disk. *AIP Advances*, *5*(4). https://doi.org/10.1063/1.4917459

Hussain, M., Rasool, M. and Mehmood, A., 2022. Radiative flow of viscous nanofluid over permeable stretched swirling disk with generalized slip. *Scientific Reports*, *12*(1), p.11038.

Kalyani, C., Kishan, N. and Reddy, M., 2017. MHD stagnation point flow of a nanofluid towards a radially stretching convectively heated disk with viscous dissipation. *Journal of Nanofluids*, *6*(1), pp.182–188.

Khan, M., Ahmed, J. and Ahmad, L., 2018. Chemically reactive and radiative von Kármán swirling flow due to a rotating disk. *Applied Mathematics and Mechanics*, *39*, pp.1295–1310.

Khan, N.A., Aziz, S. and Ullah, S., 2017. Entropy generation on MHD flow of Powell-Eyring fluid between radially stretching rotating disk with diffusion-thermo and thermo-diffusion effects. *Acta Mechanica et Automatica*, *11*(1), pp.20–32.

Khan, U., Waini, I., Ishak, A. and Pop, I., 2021. Unsteady hybrid nanofluid flow over a radially permeable shrinking/stretching surface. *Journal of Molecular Liquids*, *331*, p.115752.

Khashi'ie, N.S., Wahid, N.S., Md Arifin, N. and Pop, I., 2022. MHD stagnation-point flow of hybrid nanofluid with convective heated shrinking disk, viscous dissipation and Joule heating effects. *Neural Computing and Applications*, *34*(20), pp.17601–17613.

Mahyuddin, A.A., Lok, Y.Y. and Ahmad, S., 2022. Bödewadt flow and heat transfer in nanofluid over a permeable and radially stretching disk. *Sains Malays*, *51*, pp.619–632.

Makinde, O.D., Mabood, F., Khan, W.A. and Tshehla, M.S., 2016. MHD flow of a variable viscosity nanofluid over a radially stretching convective surface with radiative heat. *Journal of Molecular Liquids*, *219*, pp.624–630.

Mandal, S. and Shit, G.C., 2021. Entropy analysis on unsteady MHD biviscosity nanofluid flow with convective heat transfer in a permeable radiative stretchable rotating disk. *Chinese Journal of Physics*, *74*, pp.239–255.

Naganthran, K., Nazar, R. and Pop, I., 2019. Effects of heat generation/absorption in the Jeffrey fluid past a permeable stretching/shrinking disc. *Journal of the Brazilian Society of Mechanical Sciences and Engineering*, *41*, pp.1–12.

Naqvi, S.M.R.S., Muhammad, T., Saleem, S. and Kim, H.M., 2019. Significance of non-uniform heat generation/absorption in hydromagnetic flow of nanofluid due to stretching/shrinking disk. *Physica A: Statistical Mechanics and its Applications*, *553*, p.123970.

Oztop, H.F., Abu-Nada, E., 2008. Numerical study of natural convection in partially heated rectangular enclosures filled with nanofluids. *International Journal of Heat and Fluid Flow*, *29*, pp. 1326–1336.

Prasad, K.V., Vaidya, H., Makinde, O.D., Vajravelu, K. and Ramajini, V., 2020. Impact of suction/injection and heat transfer on unsteady MHD flow over a stretchable rotating disk. *Latin American Applied Research-An International Journal*, *50*(3), pp.159–165.

Ramzan, M., Riasat, S., Zhang, Y., Nisar, K.S., Badruddin, I.A., Ahammad, N.A. and Ghazwani, H.A.S., 2021. Significance low oscillating magnetic field and Hall current in the nano-ferrofluid flow past a rotating stretchable disk. *Scientific Reports*, *11*(1), p.23204.

Rashad, M.S., Liu, H. and Khan, S.A., 2021. Numerical simulation for bioconvection of unsteady stagnation point flow of Oldroyd-B nanofluid with activation energy and temperature-based thermal conductivity past a stretching disk. *CMES-Computer Modeling in Engineering and Sciences*, *130*(1), pp.233–254.

Rashidi, M.M., Ali, M., Freidoonimehr, N. and Nazari, F., 2013. Parametric analysis and optimization of entropy generation in unsteady MHD flow over a stretching rotating disk using artificial neural network and particle swarm optimization algorithm. *Energy*, *55*, pp.497–510.

Reddy, M.G., Kumar, N., Prasannakumara, B.C., Rudraswamy, N.G. and Kumar, K.G., 2021. Magnetohydrodynamic flow and heat transfer of a hybrid nanofluid over a rotating disk by considering Arrhenius energy. *Communications in Theoretical Physics*, *73*(4), p.045002.

Roşca, N.C. and Pop, I., 2017. Axisymmetric rotational stagnation point flow impinging radially a permeable stretching/shrinking surface in a nanofluid using Tiwari and Das model. *Scientific Reports*, *7*(1), p.40299.

Sarkar, G.M. and Sahoo, B., 2021. On dual solutions of the unsteady MHD flow on a stretchable rotating disk with heat transfer and a linear temporal stability analysis. *European Journal of Mechanics-B/Fluids*, *85*, pp.149–157.

Shateyi, S. and Makinde, O.D., 2013. Hydromagnetic stagnation-point flow towards a radially stretching convectively heated disk. *Mathematical Problems in Engineering*, *2013*(1), p.616947.

Soid, S.K., Ishak, A. and Pop, I., 2017. MHD flow and heat transfer over a radially stretching/shrinking disk. *Chinese Journal of Physics*, *56*(1), pp.58–66.

Takabi, B. and Salehi, S., 2014. Augmentation of the heat transfer performance of a sinusoidal corrugated enclosure by employing hybrid nanofluid. *Advances in Mechanical Engineering*, 6, p.147059.

Tulu, A., 2023. Analysis of magnetohydrodynamic micropolar nanofluid flow due to radially stretchable rotating disk employing spectral method. *Advances in Mathematical Physics*, *2023*(1), p.5283475.

Turkyilmazoglu, M., 2018. Fluid flow and heat transfer over a rotating and vertically moving disk. *Physics of Fluids*, *30*(6). https://doi.org/10.1063/1.5037460

Waini, I., Ishak, A. and Pop, I., 2019. Unsteady flow and heat transfer past a stretching/shrinking sheet in a hybrid nanofluid. *International Journal of Heat and Mass Transfer*, *136*, pp.288–297.

Waini, I., Ishak, A. and Pop, I., 2021. Agrawal flow of a hybrid nanofluid over a shrinking disk. *Case Studies in Thermal Engineering*, *25*, p.100950.

Weidman, P., 2014. Axisymmetric stagnation-point flow on a spiraling disk. *Physics of Fluids*, *26*(7). https://doi.org/10.1063/1.4887260

6 Nanofluid Flow, Heat, and Mass Transfer at a Stretching Cylinder

6.1 NANOFLUID FLOW PAST A STRETCHING CYLINDER

As mentioned in the abstract, heat transfer is a process that occurs in many practical and applied applications (Kumar et al., 2021). To improve heat transfer efficiency, innovative materials are becoming used when traditional materials and methods are unable to meet thermal requirements. Using numerous methods, researchers are able to characterize the thermal and hydraulic performance of various fluids so that their use can be optimized. One group of fluids is characterized as nanofluids; this generally refers to fluids with suspension of small (< 100nm) in the otherwise normal liquids (Sheikholeslami, 2015). In many industrial processes that involve heat transfer, nanofluids are used because they tend to have excellent heat transfer capabilities on account of their elevated thermal conductivity. Microchip cooling, coolant, lubrication in automobiles, and geothermal power extraction are a few representative examples of nanofluid use (see the discussion in Sheikholeslami, 2015). In some instances, thermal requirements can only be fulfilled by nanofluids (Khan et al., 2019). However, nanofluids generally possess a larger viscosity, which reduces pumping efficiency. As a result, research is ongoing to balance the improvements to heat transfer with the penalty from viscosity (Mukhopadhyay et al., 2013).

Motivated by the widespread applications of nanofluid flow caused by stretched cylinders, many researchers have this physical situation. For example, investigated the effects of Newtonian heating on nanofluid flow past a cylinder when suction/blowing occurs. Nourazar et al. (2017) scrutinized the flow of a nanofluid over a stretched cylinder and reported the effects of suction/blowing. Using a Range-Kutta Fehlberg numerical scheme, Khan et al. (2017a) explored the combined influences of radiation, thermal slip, and suction/blowing in the boundary layer of nanofluid flow over a nonlinearly stretched cylinder in the presence of a magnetic field. They reported enhancements in fluid temperatures and Biot numbers. Using Buongiorno's model, Salahuddin et al. (2017) examined the effects of generalized slip on Carreau nanofluid flow over a stretching cylinder in the presence of a chemical reaction. They showed the diminishing concentration and magnitude of velocity due to an increase in the chemical reaction parameter. Using a Runge-Kutta-Fehlberg method, Khan et al. (2017b) obtained numerical solutions with Joule heating and convective

DOI: 10.1201/9781003571278-6

boundary conditions on Carreau nanofluid flow due to a stretched cylinder in the presence of a magnetic field. They showed that for large values of the Weissenberg number, the velocity of shear-thinning fluids was improved, but the reverse was noted for shear-thickening fluids. Azam et al. (2017) discussed the influences of nonlinear thermal radiation of MHD unsteady Carreau nanofluid flow past a stretched cylinder near a stagnation point. They also reported the impact of Brownian motion and thermophoresis.

Pandey and Kumar (2017) explored free convective nanofluid flow due to a stretching cylinder considering viscous dissipation, a heat sink, thermal radiation, and slip boundary conditions. They observed an increase in temperature for smaller values of the natural convection parameter, and a rise in the rate of heat transfer was noted as porosity decreased. Considering Dufour and Soret effects, Jagan et al. (2018) analyzed Jeffrey nanofluid flow past a stretching cylinder in the presence of a magnetic field, and they reported a rise in solute boundary layer thickness but a reduction in mass transfer rate as the Soret number increased.

Gholinia et al. (2019) quantified the influence of thermal radiation on hybrid nanofluid flow past a porous stretched cylinder. Maskeen et al. (2019) investigated hybrid nanofluid flow over a stretchy cylinder considering copper and alumina nanoparticles. Waini et al. (2020) examined hybrid nanofluid flow due to a shrinking/stretching cylinder near a stagnation point, and they showed an enhancement in the rate of heat transfer. Taking account of viscous dissipation and Joule heating, MHD nanofluid flow over a stretching cylinder was examined by Mishra and Kumar (2020); they reported the effects of velocity and thermal slip on flow and heat transport. Using a mathematical model, Walelign et al. (2020) analyzed Casson nanofluid flow over an inclined stretching cylinder in the presence of a magnetic field. They used the Homotopy Analysis Method (HAM) to obtain analytic solutions for the problem. They noted an improved rate of mass transfer for rising values of the curvature and buoyancy parameters. Singh et al. (2021) reported the effects of melting heat transfer on magnetic nanofluid flow over a stretched cylinder. Kumar et al. (2021a) explored magnetized nanofluid flow across a stretching cylinder. With the help of an improved Fourier thermal flux equation, Kumar et al. (2021b) inspected hybrid nanofluid flow past a stretching cylinder. Ali et al. (2021) investigated MHD flow of a hybrid nanofluid over a stretching cylinder and discussed the combined effects of nonlinear thermal radiation and non-uniform heat flux. Waqas et al. (2021) examined MHD hybrid nanofluid flow over a vertical stretching cylinder and found that thermal transport increases as the magnetic number rises. Sreenivasa et al. (2022) inspected the flow of binary nanofluids over a stretchable cylinder in a porous medium. Salmi et al. (2022) examined hybrid nanofluid flow across a stretchy cylinder in the presence of two-phase chemical reactions. Siddiqui et al. (2022) explored the three-dimensional flow of a nanofluid past a stretching cylinder and calculated the entropy generation for such flows. Gul et al. (2023) examined ternary hybrid nanofluid flow over a stretching cylinder in the presence of a magnetic field. They reported on the influences of homogeneous-heterogeneous chemical reactions on mass transfer.

Considering activation energy and variable thermal conductivity, Paul et al. (2023) discussed Casson-Maxwell hybrid nanofluid flow due to an unsteady, stretchy cylinder. They observed a 30% rise in absolute skin friction for a hybrid nanofluid compared to a Casson-Maxwell nanofluid model. Considering Cattaneo–Christov heat flux, Ashraf et al. (2023) analyzed a Jeffery nanofluid flow across a stretching cylinder in the presence of a magnetic field. Paul et al. (2024) investigated Casson ternary hybrid nanofluid flow over an exponentially stretched cylinder and reported the effects of a magnetic field on flow, heat, and mass transfer. Moreover, their study showed that Cattaneo-Christov heat flux played a significant role in heat transfer. Compared to a Casson nanofluid, they observed an 11% increase in skin friction with a hybrid nanofluid. Taking viscous dissipation into consideration, Mishra et al. (2024) explored a blood-based ferrofluid flow along a stretching cylinder. Sulaiman Basha and Anthony (2024) explored the combined influences of an inclined magnetic field and nonlinear thermal radiation on hybrid nanofluid flow due to a stretched cylinder. Considering Arrhenius activation energy and a porous medium, Ur Rehman et al. (2024) examined the combined consequences of a magnetic field and thermal radiation on nanofluid flow due to a linearly stretched cylinder. Taking non-linear thermal radiation and Cattaneo-Christov heat flux, Vijatha and Bala Anki Reddy (2024) analyzed the flow of an Eyring-Powell hybrid nanofluid over a cylinder. Using a Cattaneo-Christov double diffusion model, Rath and Nayak (2024) reported the effects of a magnetic field on a Casson hybrid nanofluid flow over a stretchy cylinder.

Almeshaal et al. (2024) discussed double diffusive nanofluid flow past an elongated cylinder and highlighted its applications in energy storage systems. Considering activation energy and chemical reaction, Zhao et al. (2024) analyzed the bioconvective buoyancy-driven flow of a Casson nanofluid due to a vertically stretched cylinder in a Darcy–Forchheimer porous medium. Considering thermal stratification, Naveed Khan et al. (2024) examined the convective flow of tangent hyperbolic nanofluid across a stretching cylinder and reported on the effects of suction/blowing. With a Cattaneo-Christov heat flux, the effects of Darcy-Forchheimer porous medium on Carreau hybrid nanofluid flow due to a stretching cylinder were analyzed by Sarma and Paul (2024).

It is clear from the above discussion that the analysis of flow and heat transfer phenomena over a stretching cylinder has obtained a lot of attention within the research community. This interest reflects the myriad of physical processes that can occur in nanofluid flows as well as the widespread industrial applications. In this context, the present study will explore many aspects of heat transfer in Cu/Ag-water nanofluid flow past an elongated cylinder.

6.2 MATHEMATICAL ANALYSIS OF THE PROBLEM

Consider a Newtonian nanofluid along a stretching cylinder with a uniform surface temperature, as shown in Figure 6.1. The stretching surface of the cylinder causes an incompressible laminar boundary layer nanofluid flow towards the x-direction, as illustrated in the figure. The x-axis is measured along the axis of the cylindrical tube, and the r-axis is measured in the radial direction.

The continuity, momentum, and energy equations governing such type of flow and the associated boundary conditions can be written as

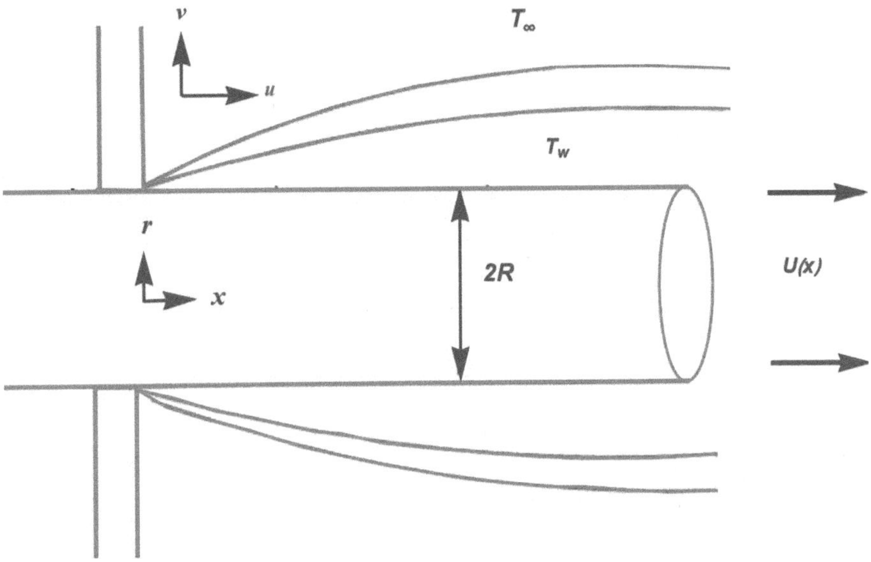

FIGURE 6.1 Illustration of fluid flowing along a stretching cylinder.

$$\frac{\partial(ru)}{\partial x}+\frac{\partial(r\upsilon)}{\partial r}=0, \tag{6.1}$$

$$u\frac{\partial u}{\partial x}+\upsilon\frac{\partial u}{\partial r}=\frac{\nu_{nf}}{r}\frac{\partial}{\partial r}\left(r\frac{\partial u}{\partial r}\right), \tag{6.2}$$

$$u\frac{\partial T}{\partial x}+\upsilon\frac{\partial T}{\partial r}=\frac{k_{nf}}{(\rho c_p)_{nf}}\frac{1}{r}\frac{\partial}{\partial r}\left(r\frac{\partial T}{\partial r}\right), \tag{6.3}$$

where u and υ are the components of velocity respectively in the x and r directions, $\nu_{nf}=\mu_{nf}/\rho_{nf}$ is the kinematic viscosity of nanofluid, ρ_{nf} is the density of nanofluid, μ_{nf} is the coefficient of viscosity of nanofluid, k_{nf} represents the effective thermal conductivity, $(\rho c_p)_{nf}$ stands for the effective heat capacitance of nanofluid, and T is the fluid temperature.

The appropriate boundary conditions for the problem are given by

$$u=U(x),\upsilon=0,T=T_w,\quad at\quad r=R, \tag{6.4a}$$

$$u\to 0,T\to T_\infty\quad as\quad r\to\infty. \tag{6.4b}$$

Here, $U(x) = U_0\, x/L$ is the stretching velocity, T_w is the uniform surface temperature, U_0 is the reference velocity, T_∞ is the ambient temperature, and L is the characteristic length. The effective fluid properties are given by Konai and Mukhopadhyay (2024) as:

$$\mu_{nf} = \frac{\mu_f}{(1-\phi)^{2.5}}, \tag{6.5a}$$

$$\rho_{nf} = (1-\phi)\rho_f + \phi\rho_s, \tag{6.5b}$$

$$\frac{k_{nf}}{k_f} = \frac{(k_s + 2k_f) - 2\phi(k_f - k_s)}{k_s + 2k_f + \phi(k_f - k_s)}, \tag{6.5c}$$

$$(\rho c_p)_{nf} = (1-\phi)(\rho c_p)_f + \phi(\rho c_p)_s. \tag{6.5d}$$

Here, φ is the volume fraction of the nanoparticles. The subscripts f, s, and nf refer to the base fluid, solid nanoparticles, and the nanofluid, respectively.

The thermophysical properties of water and Cu, Ag nanoparticles are listed in Table 6.1 (from Parveen and Mahapatra, 2022).

6.3 THE SOLUTION PROCEDURE

The continuity equation is automatically satisfied by the introduction of the stream function ψ as

$$u = \frac{1}{r}\frac{\partial \psi}{\partial r}, \upsilon = -\frac{1}{r}\frac{\partial \psi}{\partial x}. \tag{6.6a}$$

TABLE 6.1
Thermophysical properties of water, Cu, and Ag nanoparticles

Properties	Water	Cu	Ag
ρ (kg/m^3)	997.1	8933	10500
k (W/m°C)	0.613	401	429
c_p (J/kg°C)	4179	385	235

Source: Parveen and Mahapatra (2022).

The following similarity variables are introduced:

$$\eta = \frac{r^2 - R^2}{2R}\left(\frac{U}{\nu_f x}\right)^{\frac{1}{2}}, \tag{6.6b}$$

$$\psi = (U\nu_f x)^{\frac{1}{2}} Rf(\eta), \tag{6.6c}$$

$$\theta(\eta) = \frac{T - T_\infty}{T_w - T_\infty} \tag{6.6d}$$

and upon substitution of Eq. (6.6d) into Eqs. (6.2), (6.3), (6.4a), and (6.4b), the governing equations and the boundary conditions reduce to:

$$\frac{\frac{\mu_{nf}}{\mu_f}}{\frac{\rho_{nf}}{\rho_f}}(1+2\alpha\eta)f''' + 2\frac{\frac{\mu_{nf}}{\mu_f}}{\frac{\rho_{nf}}{\rho_f}}\alpha f'' + ff'' - f'^2 = 0, \tag{6.7}$$

$$\frac{\frac{\kappa_{nf}}{\kappa_f}}{\frac{(\rho c_p)_{nf}}{(\rho c_p)_f}}\left\{(1+2\alpha\eta)\theta'' + 2\alpha\theta'\right\} + Pr\, f\theta' = 0, \tag{6.8}$$

$$f' = 1, f = 0, \theta = 1 \text{ at } \eta = 0, \tag{6.9a}$$

and

$$f' \to 0, \theta \to 0 \text{ as } \eta \to \infty, \tag{6.9b}$$

where the prime denotes differentiation with respect to η and $\alpha = (\nu_f L/U_0 R^2)^{1/2}$ is the curvature parameter.

If $\alpha = 0$ (i.e., $R \to \infty$), the problem under consideration (with $\varphi = 0$) reduces to the boundary layer flow along a stretching flat plate considered by Ali (1994), with $m = 1$. Moreover, when $\alpha = 0$ (stretching flat plate) subjected to Eq. (6.9) with $\varphi = 0$ (for an ordinary viscous fluid), the analytical solutions of Eqs. (6.7) and (6.8) are given by Crane (1970) and Grubka and Bobba (1985), respectively. The closed-form solution of Eq. (6.7) with $\alpha = 0, \varphi = 0$ is $f(\eta) = 1 - e^{-\eta}$.

6.4 NUMERICAL RESULTS AND DISCUSSION

To analyze the results, numerical computations are presented for viscous fluid flow ($\varphi = 0$) past a stretching flat plate (for $\alpha = 0$ (i.e., $R \to \infty$)), and the results are presented in Table 6.2.

Moreover, to examine the accuracy of the numerical method used in this study, $[-\theta'(0)]$ for ordinary viscous fluid flow past a stretching plate having a uniform surface, the temperature is compared with the existing results of Grubka and Bobba (1985), Ali (1994), Ishak and Nazar (2009), and Mukhopadhyay et al. (2012) and the comparison is presented in Table 6.3 which shows near-perfect agreement.

In Figure 6.2a, horizontal velocity profiles are shown for different values of nanoparticle volume fraction φ; the rate of transport diminishes with increasing distance (η) along the stretched cylinder. The horizontal velocity curves also show that fluid velocity decreases with the rise in nanoparticle volume fraction φ. The frictional force increases with the rise in nanoparticle volume fraction φ , and the velocity diminishes. In all cases, the fluid velocity is found to vanish at some distance from the surface of the cylinder (at η =15). Compared to Ag-water nanofluid, a higher velocity is observed for Cu-water nanofluid. On the other hand, a rise in temperature is noted for rising nanoparticle volume fraction, φ. Since the fluid velocity decreases for rising values of nanoparticle volume fraction, the temperature rises as a consequence. Compared to Cu-water nanofluid, higher temperatures are observed for the Ag-water nanofluid.

The fluid velocity increases with increasing values of α (see Figure 6.3a), which agrees with the findings of Mukhopadhyay (2013). A larger velocity is observed for the Cu-water nanofluid compared to that for Ag-water nanofluid. The temperature also rises in this case. The thermal boundary layer thickness diminishes with an increase in curvature parameter α. Compared to Cu nanoparticles, a slightly higher temperature is observed for Ag-water nanofluid. The curvature parameter α has a

TABLE 6.2
Values of ($f''(0)$) obtained from analytical and numerical solutions

Analytical solution	Numerical solution	Error
-1.00000	-1.00001	0.00001

TABLE 6.3
Values of ($-\theta'(0)$) for ordinary viscous fluid flow past a flat plate with uniform surface temperature

Grubka and Bobba (1985)	Ali (1994)	Ishak and Nazar (2009)	Mukhopadhyay et al. (2012) ($\lambda = B = M = 0$, $Pr = 1$)	Present results ($\alpha = 0$, $Pr = 1$)
0.5820	0.5801	0.5820	0.5821	0.5820

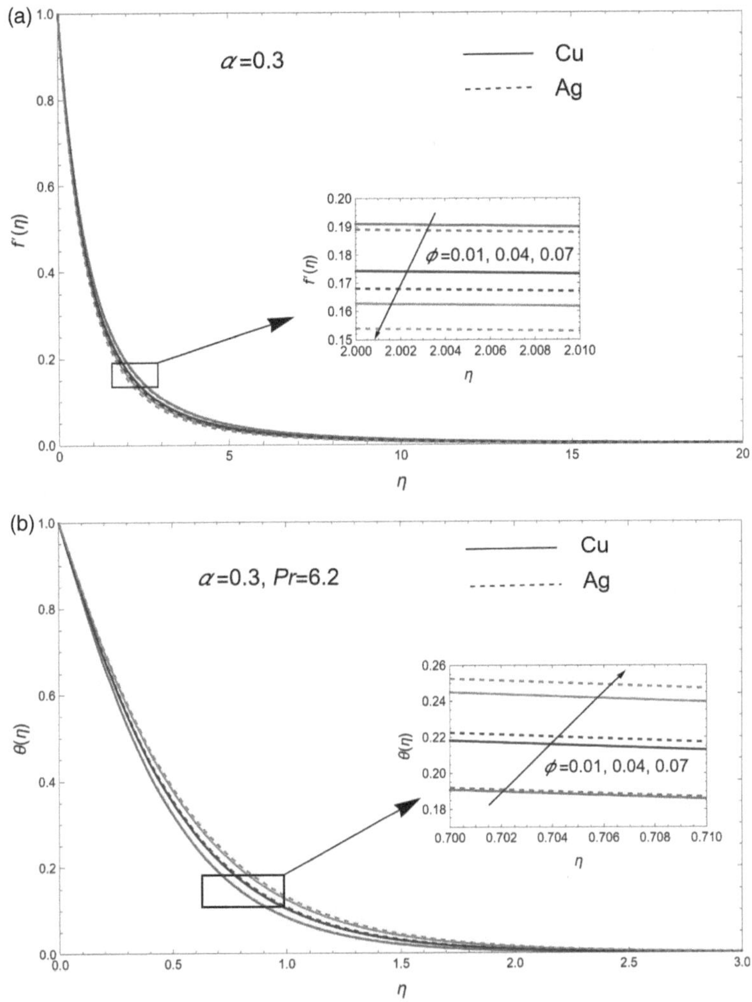

FIGURE 6.2 Effects of φ on (a) velocity $f'(\eta)$ and (b) temperature $\theta(\eta)$.

substantial effect on fluid velocity and temperature since the lateral surface of the cylinder changes with the curvature parameter.

The temperature of the nanofluid reduces with the rise in Prandtl number Pr, which is evident from Figure 6.4.

The magnitude of the velocity gradient at the surface increases with the curvature parameter α; larger velocity gradients produce larger skin friction. Thus, the magnitude of the skin friction increases with increasing curvature parameter α as well as with nanoparticle volume fraction φ. From Table 6.4, it is found that shear stress at the wall is negative for all values of α and φ. A negative sign of $f''(0)$ corresponds to a dragging force on the fluid by the surface of the cylinder, which is not surprising

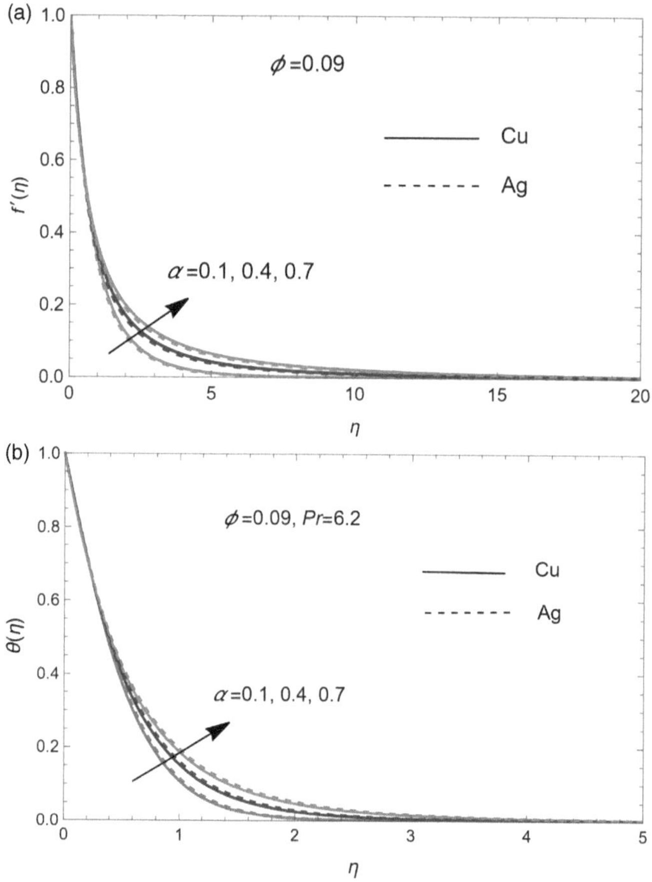

FIGURE 6.3 Effects of α on (a) velocity $f'(\eta)$ and (b) temperature $\theta(\eta)$.

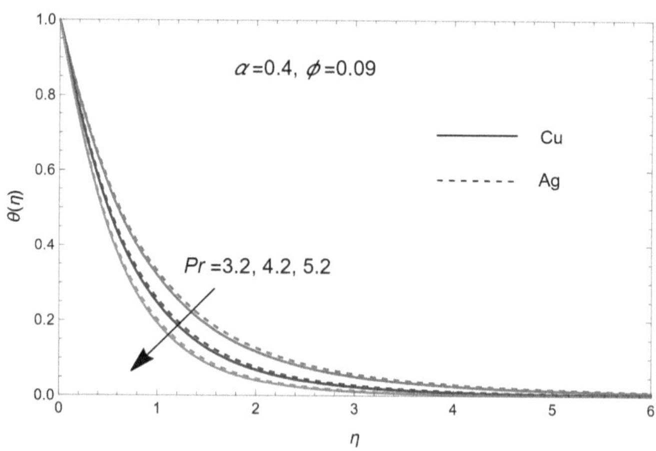

FIGURE 6.4 Effects of Pr on temperature $\theta(\eta)$.

TABLE 6.4
Values of $f''(0)$ and $\theta'(0)$ for Cu-water nanofluid with $Pr = 6.2$

α	ϕ	$f''(0)$	$\theta'(0)$
0	0	-1.00001	-1.77094
0.2		-1.07410	-1.83580
0.4		-1.14800	-1.89950
0.6		-1.22093	-1.96225
0	0.02	-1.04981	-1.69870
	0.04	-1.09108	-1.63117
	0.06	-1.12507	-1.56783
0.3		-1.23553	-1.66327
0.4		-1.27212	-1.69452
0.5		-1.30846	-1.72553
	0.03	-1.25526	-1.82326
	0.04	-1.27474	-1.78965
	0.05	-1.29243	-1.75709

because this flow is induced by the stretching of a cylinder. Table 6.4 also displays the nature of the heat transfer coefficient with curvature parameter α and nanoparticle volume fraction.

The magnitude of the temperature gradient at the surface increases as the curvature parameter α increases, and consequently, the rate of heat transfer increases as well. As a result, the Nusselt number ($Nu = -\theta'(0)/\sqrt{\text{Re}}$) increases with the curvature parameter α (Table 6.4). However, the magnitude of the temperature gradient at the surface diminishes with the rise in the nanoparticle volume fraction φ (see Table 6.4). The rate of heat transport at the surface of the stretched cylinder diminishes with the rise in nanoparticle volume fraction φ.

6.5 CONCLUDING REMARKS

This chapter presents the comparative study for the flow of Cu/Ag-water nanofluid past a horizontal stretching cylinder. A uniform surface temperature has been considered in this study to explore the heat transfer characteristics clearly. The effects of the different parameters on flow and thermal behaviors have been carefully inspected and reported in detail. The curvature parameter of the stretching cylinder and the nanoparticle volume fraction both play vital roles, as the flow and temperature fields are significantly influenced by these parameters. With the rise in curvature parameter of the stretching cylinder, velocity gradient as well as temperature gradient at the surface both are found to reduce.

The present chapter thus contributes to the understanding of the behavior of nanofluid flow and heat transfer past a stretching cylinder. It highlights nanofluid optimization of heat transport mechanisms in different applications.

Nomenclature

c_p	specific heat (J/kg K)
Pr	Prandtl number
T	temperature of the liquid (K)
T_∞	free stream temperature (K)
T_w	surface temperature (K)
u, υ	velocity component in x, y direction (m/s)
α	curvature parameter
φ	nanoparticle volume fraction
η	similarity variable
κ	liquid thermal conductivity ($Wm^{-1}K^{-1}$)
ν	kinematic viscosity (m^2/s)
θ	non-dimensional fluid temperature
ρ	liquid density ($kg\ m^{-3}$)
ψ	stream function

REFERENCES

Ali, A., Kanwal, T., Awais, M., Shah, Z., Kumam, P. and Thounthong, P., 2021. Impact of thermal radiation and non-uniform heat flux on MHD hybrid nanofluid along a stretching cylinder. *Scientific Reports*, *11*(1), p.20262.

Ali, M.E., 1994. Heat transfer characteristics of a continuous stretching surface. *Wärme-und Stoffübertragung*, *29*(4), pp.227–234.

Almeshaal, M.A., Majeed, A., Ijaz, N., Alvi, H., Palaniappan, M., Choubani, K., Khan, S.U. and Kolsi, L., 2025. Numerical simulations for thermally developed double diffusion flow of nanoparticles due to elongated cylinder with multiple slip effects: Applications to energy storage system. *Numerical Heat Transfer, Part A: Applications*, *86*, pp.4651–4666.

Ashraf, M.B., Tanveer, A. and Ulhaq, S., 2023. Effects of Cattaneo-Christov heat flux on MHD Jeffery nano fluid flow past a stretching cylinder. *Journal of Magnetism and Magnetic Materials*, *565*, p.170154.

Azam, M., Khan, M. and Alshomrani, A.S., 2017. Unsteady radiative stagnation point flow of MHD Carreau nanofluid over expanding/contracting cylinder. *International Journal of Mechanical Sciences*, *130*, pp.64–73.

Crane, L.J., 1970. Flow past a stretching plate. *Zeitschrift für angewandte Mathematik und Physik ZAMP*, *21*, pp.645–647.

Gholinia, M., Armin, M., Ranjbar, A.A. and Ganji, D.D., 2019. Numerical thermal study on CNTs/C2H6O2–H2O hybrid base nanofluid upon a porous stretching cylinder under impact of magnetic source. *Case Studies in Thermal Engineering*, *14*, p.100490.

Grubka, L.J. and Bobba, K.M., 1985. Heat transfer characteristics of a continuous stretching surface with variable temperature. *Journal of Heat Transfer*, *107*(1), pp.248–250.

Gul, H., Ramzan, M., Saleel, C.A. and Kadry, S., 2025. A comparative analysis of ternary-hybrid nanofluid flows through a stretching cylinder influenced by an induced magnetic field with homogeneous–heterogeneous reactions. *Numerical Heat Transfer, Part A: Applications*, *86*, pp.1–16.

Hayat, T., Khan, M.I., Waqas, M. and Alsaedi, A., 2017. Newtonian heating effect in nanofluid flow by a permeable cylinder. *Results in Physics*, *7*, pp.256–262.

Ishak, A. and Nazar, R., 2009. Laminar boundary layer flow along a stretching cylinder. *European Journal of Scientific Research*, *36*(1), pp.22–29.

Jagan, K., Sivasankaran, S., Bhuvaneswari, M., Rajan, S. and Makinde, O.D., 2018, November. Soret and Dufour effect on MHD Jeffrey nanofluid flow towards a stretching cylinder with triple stratification, radiation and slip. In *Defect and Diffusion Forum* (Vol. 387, pp. 523–533). Trans Tech Publications Ltd.

Khan, I., Malik, M.Y., Hussain, A. and Khan, M., 2017b. Magnetohydrodynamics Carreau nanofluid flow over an inclined convective heated stretching cylinder with Joule heating. *Results in Physics*, *7*, pp.4001–4012.

Khan, M., Irfan, M. and Khan, W.A., 2019. Heat transfer enhancement for Maxwell nanofluid flow subject to convective heat transport. *Pramana*, *92*(2), p.17.

Khan, M.I., Tamoor, M., Hayat, T. and Alsaedi, A., 2017a. MHD boundary layer thermal slip flow by nonlinearly stretching cylinder with suction/blowing and radiation. *Results in Physics*, *7*, pp.1207–1211.

Konai, S. and Mukhopadhyay, S., 2024. Insight into the forced convective radiative Stefan flow of nanofluid over an unsteady stretched sheet. *International Journal of Computational Materials Science and Engineering*, p.2450017.

Kumar, R.N., Gowda, R.P., Abusorrah, A.M., Mahrous, Y.M., Abu-Hamdeh, N.H., Issakhov, A., Rahimi-Gorji, M. and Prasannakumara, B.C., 2021a. Impact of magnetic dipole on ferromagnetic hybrid nanofluid flow over a stretching cylinder. *Physica Scripta*, *96*(4), p.045215.

Maskeen, M.M., Zeeshan, A., Mehmood, O.U. and Hassan, M., 2019. Heat transfer enhancement in hydromagnetic alumina–copper/water hybrid nanofluid flow over a stretching cylinder. *Journal of Thermal Analysis and Calorimetry*, *138*, pp.1127–1136.

Mishra, A. and Kumar, M., 2020. Velocity and thermal slip effects on MHD nanofluid flow past a stretching cylinder with viscous dissipation and Joule heating. *SN Applied Sciences*, *2*(8), p.1350.

Mishra, N.K., Raizah, Z., Anwar, S., Almusawa, M.Y. and Saeed, A., 2024. A theoretical analysis of the electrically conducting blood-based Ferrofluid flow through a stretching cylinder with viscous dissipation. *Numerical Heat Transfer, Part B: Fundamentals*, *85*(5), pp.538–553.

Mukhopadhyay, S., 2013. MHD boundary layer slip flow along a stretching cylinder. *Ain Shams Engineering Journal*, *4*(2), pp.317–324.

Mukhopadhyay, S., De, P.R., Bhattacharyya, K. and Layek, G.C., 2012. Slip effects on mixed convection flow along a stretching cylinder. *International Journal of Heat and Technology*, *30*(2), pp.19–24.

Mukhopadhyay, S., Mondal, I.C. and Gorla, R.S.R., 2013. Effects of partial slip on boundary layer flow and heat transfer past a stretching circular cylinder in a nanofluid. *International Journal of Microscale and Nanoscale Thermal and Fluid Transport Phenomena*, *4*(2), p.115.

Naveed Khan, M., Alhowaity, A., Wang, Z., Gepreel, K.A. and Hussien, M., (in press). Numerical analysis of the heat transfer application on a convective tangent hyperbolic nanofluid flow over a porous stretching cylinder with stratification effects. *Numerical Heat Transfer, Part A: Applications*, doi: 10.1080/10407782.2024.2353881.

Nourazar, S.S., Hatami, M., Ganji, D.D. and Khazayinejad, M., 2017. Thermal-flow boundary layer analysis of nanofluid over a porous stretching cylinder under the magnetic field effect. *Powder Technology*, *317*, pp.310–319.

Pandey, A.K. and Kumar, M., 2017. Natural convection and thermal radiation influence on nanofluid flow over a stretching cylinder in a porous medium with viscous dissipation. *Alexandria Engineering Journal*, *56*(1), pp.55–62.

Parveen, R. and Mahapatra, T.R., 2022. Heat and mass source effect on MHD double-diffusive mixed convection and entropy generation in a curved enclosure filled with nanofluid. *Nonlinear Analysis: Modelling and Control*, *27*(2), pp.308–330.

Paul, A., Sarma, N. and Patgiri, B., 2023. Thermal and mass transfer analysis of Casson-Maxwell hybrid nanofluids through an unsteady horizontal cylinder with variable thermal conductivity and Arrhenius activation energy. *Numerical Heat Transfer, Part A: Applications*, *86*, pp.2954–2979..

Paul, A., Sarma, N. and Patgiri, B., 2024. Numerical assessment of MHD thermo-mass flow of Casson ternary hybrid nanofluid around an exponentially stretching cylinder. *BioNanoScience*, *14*(5), pp.5230–5245.

Rath, C. and Nayak, A., (in press). Magnetohydrodynamic Casson SWCNT-MWCNT/EG hybrid nanofluid flow over a stretching cylinder with Cattaneo-Christov double diffusion model. *Numerical Heat Transfer, Part A: Applications*, doi: 10.1080/10407782.2024.2353355..

Salahuddin, T., Hussain, A., Malik, M.Y., Awais, M. and Khan, M., 2017. Carreau nanofluid impinging over a stretching cylinder with generalized slip effects: Using finite difference scheme. *Results in Physics*, *7*, pp.3090–3099.

Salmi, A., Madkhali, H.A., Nawaz, M., Alharbi, S.O. and Malik, M.Y., 2022. Non-Fourier modeling and numerical simulations on heat and transfer in tangent hyperbolic nanofluid subjected to chemical reactions. *International Communications in Heat and Mass Transfer*, *134*, p.105996.

Sarma, N. and Paul, A., (in press). Cattaneo-Christov heat flux effect on Darcy-Forchheimer dual-phase dusty shear-thickening carreau hybrid nanofluid flow along a stretched vertical cylinder. *Numerical Heat Transfer, Part B: Fundamentals*, doi: 10.1080/10407790.2024.2364786.

Sheikholeslami, M., 2015. Effect of uniform suction on nanofluid flow and heat transfer over a cylinder. *Journal of the Brazilian Society of Mechanical Sciences and Engineering*, *37*, pp.1623–1633.

Siddiqui, B.K., Batool, S., mahmood ul Hassan, Q. and Malik, M.Y., 2022. Repercussions of homogeneous and heterogeneous reactions of 3D flow of Cu-water and AL2O3-water nanofluid and entropy generation estimation along stretching cylinder. *Ain Shams Engineering Journal*, *13*(1), p.101493.

Singh, K., Pandey, A.K. and Kumar, M., 2021. Melting heat transfer assessment on magnetic nanofluid flow past a porous stretching cylinder. *Journal of the Egyptian Mathematical Society*, *29*(1), p.1.

Sreenivasa, B.R., J. Faqeeh, A., Alsaiari, A., Alzahrani, H.A. and Malik, M.Y., 2022. Numerical study of heat transfer mechanism in the flow of ferromagnetic hybrid nanofluid over a stretching cylinder. *Waves in Random and Complex Media*, pp.1–17.

Sulaiman Basha, M.I. and Anthony, D.M.G., 2024. Numerical investigation of non-linear radiative flow of hybrid nanofluid past a stretching cylinder with inclined magnetic field. *Numerical Heat Transfer, Part B: Fundamentals*, *85*(7), pp.842–866.

Ur Rehman, M.A., Farooq, M.A. and Irfan, M., (in press). Numerical analysis of magneto-hydrodynamic free stream and radiative heat transfer flow of nanofluid in a permeable medium over a linearly stretched cylinder with Arrhenius activation energy. *Numerical Heat Transfer, Part B: Fundamentals*, doi: 10.1080/10407790.2024.2362955..

Vijatha, M. and Bala Anki Reddy, P., 2024. Second law analysis on magnetohydrodynamics flow of Eyring-Powell hybrid nanofluid over a cylinder/plate with non-linear thermal radiation and Cattaneo-Christov heat flux: Numerical and semi-analytical simulations. *Numerical Heat Transfer, Part B: Fundamentals*, 86, pp. 1307-1335..

Waini, I., Ishak, A. and Pop, I., 2020. Hybrid nanofluid flow towards a stagnation point on a stretching/shrinking cylinder. *Scientific Reports*, *10*(1), p.9296.

Walelign, T., Haile, E., Kebede, T. and Walelgn, A., 2020. Analytical study of heat and mass transfer in MHD flow of chemically reactive and thermally radiative Casson nanofluid over an inclined stretching cylinder. *Journal of Physics Communications*, *4*(12), p.125003.

Waqas, H., Naqvi, S.M.R.S., Alqarni, M.S. and Muhammad, T., 2021. Thermal transport in magnetized flow of hybrid nanofluids over a vertical stretching cylinder. *Case Studies in Thermal Engineering*, *27*, p.101219.

Zhao, X., Leng, Y., Ramzan, M., Mahmood, S., Shahmir, N., Alshahrani, S. and Kadry, S., (in press). Buoyancy driven bioconvective Casson nanofluid flow over a vertical stretching cylinder in Darcy–Forchheimer permeable medium with Arrhenius activation energy and chemical reaction. *Numerical Heat Transfer, Part A: Applications*, doi: 10.1080/10407782.2024.2333502.

7 Bio-magnetic Nanofluid Fluid Flow in a Channel with Stretching Walls

7.1 INTRODUCTION

The mechanism of peristalsis involves the repetitive, wave-like contractions and relaxations of the walls of a flexible tube to convey fluids. In biological systems, this mechanism is crucial for things like moving food through the digestive tract and promoting blood flow in tiny blood arteries. It is especially useful for moving delicate or non-Newtonian fluids because it doesn't require valves to function. This idea is also used in peristaltic pumps, which are frequently used for accurate and contamination-free fluid handling in industrial, chemical, and medical contexts. The theoretical investigation of peristalsis was carried out by Latham (1966). Numerous authors have advanced peristaltic transport in a variety of media, influenced by different physical properties; we highlight the intriguing results found in Fung and Yih (1968), Srivastava and Srivastava (1984), Hariharan et al. (2008), Misra and Pandey (1999), Srivastava and Srivastava (1985), Ali et al. (2010), El-Shehawy et al. (2006), Mernone et al. (2002), and Siddiqui and Schwarz (1994).

Nanofluids are novel heat transfer fluids made by mixing nanoparticles –usually ranging in size from 1 to 100 nanometers – with a base liquid, like ethylene glycol, water, or oil. These nanoparticles, which are composed of metals, oxides, or carbon-based materials like graphene or carbon nanotubes, significantly improve the fluid's capacity for heat transmission and thermal conductivity because of their high surface area and remarkable thermal characteristics. Applications for nanofluids are numerous and include targeted medicine delivery, thermal treatment, and cooling systems for electronics, automobile engines, and power plants. Their potential to increase energy efficiency and make compact designs possibly makes them an important study topic for effective and sustainable thermal management solutions.

Nanofluids are used in many different fields, where their remarkable thermal characteristics are used to improve efficiency and performance. They enhance cooling systems in electronics, guaranteeing efficient heat dissipation for parts like CPUs and GPUs. They improve air conditioning, engine cooling, and braking efficiency in vehicle systems. Their capacity to maximize heat transmission makes renewable energy technologies like solar thermal collectors and geothermal systems advantageous, and industrial processes use them in heat exchangers for the production of

DOI: 10.1201/9781003571278-7

chemicals and power. Thermal cancer therapy and tailored drug delivery are two scientific advances made possible by nanofluids. In mixed convective peristaltic transport of nanofluid with entropy generation, Akbar et al. (2020) examined the impacts of Hall and thermal radiation. In a porous medium, Riaz et al. (2020) investigated heat transmission and Jeffrey nano-liquid peristaltic propulsion via a symmetrical conduit with moving walls. Khazayinejad et al. (2021) examined the peristaltic movement of a biological graphene-blood nanofluid in a porous medium while taking into account heat radiation and an angled magnetic field. Entropy generation and activation energy analysis in mixed convective peristaltic transport of Sutterby nanofluid were investigated by Hayat et al. (2021). A variety of other aspects of nanofluid flow were also examined by many authors (see Ajithkumar et al., 2023; Akbar et al., 2022; Ramamoorthy and Pallavarapu, 2022; Afsar et al., 2023).

In order to improve a base fluid's thermal and physical characteristics, two or more distinct kinds of nanoparticles are scattered throughout it to create a hybrid nanofluid, a novel class of engineered fluids. These nanoparticles may be composed of carbon-based materials, metals, metal oxides, or other nanostructures. Compared to ordinary nanofluids, hybrid nanofluids have better thermal conductivity, heat transfer efficiency, and stability, which makes them ideal for applications that need good thermal management. Numerous industries, including automotive (radiator cooling), electronics (chip cooling), biomedical (drug administration and treatment of hyperthermia), and renewable energy (solar thermal systems), employ them extensively. Because of their improved thermophysical characteristics, hybrid nanofluids hold promise as an energy-efficient and high-performance system solution. The effects of magnetic forces on the peristaltic transport of hybrid bio-nanofluid (*Au-Cu* nanoparticles) with a moderate Reynolds number have been studied by Abo-Elkhair et al. (2021). The peristaltic flow of a hybrid nanofluid inside an elliptic duct with sinusoidally advancing boundaries was studied by McCash et al. (2021) using entropy analysis. Iqbal et al. (2022) analyzed the entropy generation and Hall current dynamics in the peristaltic transport of a hybrid nanofluid. Using the electroosmosis phenomena, Guedri et al. (2022) addressed the Hall effects and entropy-generating applications for peristaltic flow of a modified hybrid nanofluid. Iftikhar and Sadaf (2023) modeled a modified hybrid nanofluid with convective boundary conditions and MHD in a peristaltic diverging tube. Several researchers examined various other aspects of hybrid nanofluid flow (Iqbal and Abbasi, 2024; Hayat et al., 2024; Iqbal et al., 2024; Akbar et al., 2024; Thirunavukarasan and Sucharitha, 2025).

In the context of peristaltic motion, where waves travel along flexible walls to create fluid flow, magnetohydrodynamics (MHD) in peristaltic transport studies the interplay between electrically conducting fluids and magnetic fields. Heat transfer, pressure distribution, and velocity profiles are just a few of the flow characteristics that are greatly impacted by this interaction. In biomedical engineering, MHD peristaltic transport finds extensive use in artificial pumps, blood flow modelling under magnetic fields, and magnetic drug targeting. Additionally, it has industrial uses in material processing, nuclear cooling systems, and microfluidic devices. MHD and peristaltic motion combine to provide novel approaches to regulated fluid flow in intricate systems, which makes it an intriguing topic for interdisciplinary study.

Maraj et al. (2023) examined the menthol-based nanofluid's electro-osmotically controlled magneto hydrodynamic peristaltic flow in a uniform channel with a shape factor. The impact of magnetic force and a modest Reynolds number on MHD has been examined by Alqarni et al. (2023) Jeffrey hybrid nanofluid through a peristaltic channel. Casson nanofluids with slip effects were studied by Vaidya et al. (2024) using magnetohydrodynamic peristaltic propulsion over a heterogeneous rough channel. The recent discussions related to magnetohydrodynamics and peristalsis of nanofluids are presented in Sucharitha et al. (2019), Magdy et al. (2024), Tanveer et al. (2024), and Ramki and Lakshminarayana (2024).

A substance with a solid matrix and interconnected pores that allow fluids to flow is called a porous medium. Porous media are essential in many different applications because of their special structure. They are employed in environmental research to model the movement of contaminants and filter groundwater. Porous media are essential to fuel cell design, carbon capture and storage, and increased oil recovery in the energy sector. Drug delivery systems and scaffolds for tissue engineering are examples of medical applications where the pore structure facilitates fluid movement and cell proliferation. They also play a key role in industrial processes like heat exchange, filtration, and catalysis, which facilitate effective material handling and energy transmission. Using a porous media with chemical reaction and wall traits, Abd-Alla et al. (2023) examined the heat and mass transmission in a peristaltic rotating frame Jeffrey fluid. A computational investigation using the Galerkin finite element method was conducted by Tong et al. (2022) on peristaltic blood transport in a non-Newtonian fluid contained by a porous soaked tube. Ismael et al. (2022) investigated the effects of pair stresses and thermal micropolarity on the peristaltic flow of a bi-viscous nanofluid through a porous medium. More recently, Aslam and Noreen (2024), Ijaz et al. (2024), Mekheimer et al. (2024), and Raghunath et al. (2024) have provided more fascinating studies that look at the role of the porous medium in peristaltic transport.

There are potential industrial, engineering, and biological uses for the subject of this research project. Driven by the potency of the various effects presented above, we examined the effects of heat radiation, viscous dissipation, and MHD peristaltic transport of the Casson nanofluid over a porous environment. The presence of a hybrid nanofluid increases the thermal conductivity of the fluid as compared to nanofluid and base fluid. Thus, the primary goal of this research is to analyze the unsteady magnetohydrodynamic flow of Casson hybrid nanofluid in a porous symmetric channel with a $Cu\text{-}Al_2O_3$/blood hybrid nanofluid. With the use of this model, we may examine fluid dynamics within the problem's context. It's important to understand that blood viscosity varies even when the basic fluid is "pure blood" (Abraham et al., 2008; Plourde et al., 2016). Using metallic nanoparticles combined with blood to improve thermal conductivity and enable effective heat transfer to a tumor. The temperature distribution within the treated region can also be predicted and controlled by taking into account the effects of thermal radiation. The theoretical model can be successfully applied to enhance performance in a variety of engineering and industrial contexts by carefully adapting it to the unique conditions and requirements of real-world applications and by validating and refining it through simulations and experiments.

7.2 FORMULATION OF THE PROBLEM

We have examined an incompressible, unsteady, two-dimensional, MHD flow of Casson hybrid nanofluid in a porous symmetric channel. Viscous dissipation is taken into account while analyzing heat transfer. This study takes into account both thermal radiation and an inclined magnetic field of strength B_0. The peristaltic waves are traveling along the x-axis at a speed of c. The channel walls constitute the x-axis of the cartesian coordinate system, and the y-axis is oriented orthogonally to it (see Figure 7.1). Tables 7.1 and 7.2 display the physical characteristics of nanoparticles and present the equations that describe the thermophysical properties of nanofluids and hybrid nanofluids, respectively.

The wall deformation is provided as follows (see Figure 7.1):

$$H = \bar{n}x^* + d + a\sin\frac{2\pi}{\lambda}\left(x^* - ct'\right) \tag{7.1}$$

where c, t', $\bar{n}$, d, a, and λ are the speed of the sinusoidal wave, time, the irregularity parameter, the width of the channel, the amplitude, and the wavelength, respectively.

The basic equations of flow are expressed as shown in Sucharitha et al. (2019) and Awais et al. (2020):

$$\frac{\partial u'}{\partial x^*} + \frac{\partial v'}{\partial y^*} = 0, \tag{7.2}$$

$$\rho_{hnf}\left[\frac{\partial u'}{\partial t'} + u'\frac{\partial u'}{\partial x^*} + v'\frac{\partial u'}{\partial y^*}\right] = -\frac{\partial p'}{\partial x^*} + \mu_{hnf}\left(1 + \frac{1}{\beta}\right)\nabla^2 u' - \frac{\sigma_f A_4 B_0^2}{\left(1 + \left(A_4 m\right)^2\right)}\cos\eta\left(u'\cos\eta - v'\sin\eta\right) - \frac{\mu_{hnf}}{k}u', \tag{7.3}$$

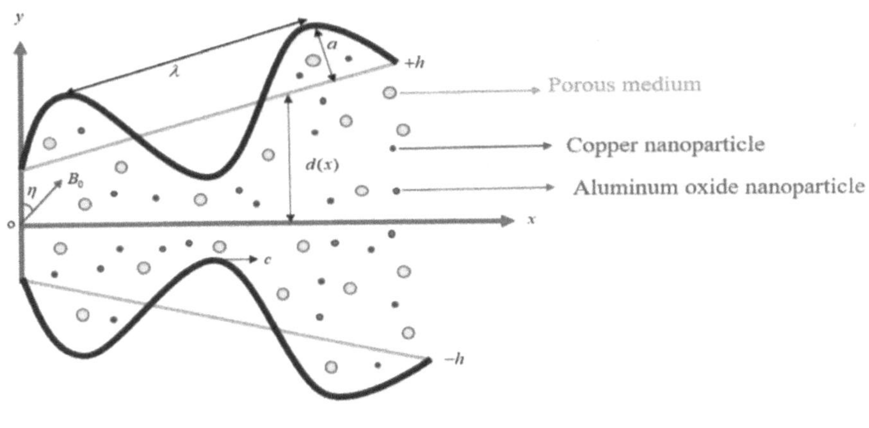

FIGURE 7.1 Geometry of the flow problem.

TABLE 7.1
Different shapes and shape factors of nanoparticles

Nanoparticle shape	Shape factor(S)	Shape
Spherical	3	
Tetrahedron	4.0613	
Cylinder	4.9	
Platelets	5.7	
Blade	8.6	
Lamina	16.1576	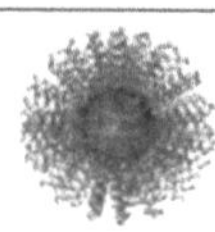

Source: Afsar et al. (2023).

$$\rho_{hnf}\left[\frac{\partial v'}{\partial t'} + u'\frac{\partial v'}{\partial x^*} + v'\frac{\partial v'}{\partial y^*}\right] = -\frac{\partial p'}{\partial y^*} + \mu_{hnf}\left(1+\frac{1}{\beta}\right)\nabla^2 v' - \frac{\sigma_f A_4 B_0^{\,2}}{\left(1+\left(A_4 m\right)^2\right)}\cos\eta\left(\mathrm{u}'\cos\eta - v'\sin\eta\right), \tag{7.4}$$

$$(\rho c)_{hnf}\left[\frac{\partial T}{\partial t'} + u'\frac{\partial T}{\partial x^*} + v'\frac{\partial T}{\partial y^*}\right] = k_{hnf}\left(\frac{\partial^2 T}{\partial x^{*2}} + \frac{\partial^2 T}{\partial y^{*2}}\right) + \mu_{hnf}\left(1+\frac{1}{\beta}\right) \left[2\left(\left(\frac{\partial u'}{\partial x^*}\right)^2 + \left(\frac{\partial v'}{\partial x^*}\right)^2\right) + \left(\frac{\partial v'}{\partial x^*} + \frac{\partial u'}{\partial y^*}\right)\right] - \frac{\partial \overline{q}_r}{\partial y^*} + Q_0. \tag{7.5}$$

The non-dimensional parameters provided are as follows:

$$
\begin{aligned}
&\overline{\psi}=\psi cd, x=\frac{x^{*}}{\lambda}, y=\frac{y^{*}}{d}, u'=uc,. v'=v\delta, p'=\frac{p\mu_{f}c\lambda}{d^{2}},\\
&\tau_{xy}=\frac{d\overline{\tau}_{xy}}{\mu c}, \overline{n}=\frac{nd}{\lambda}, t'=\frac{t\lambda}{c}, \ \varepsilon=\frac{a}{d}, \delta=\frac{d}{\lambda}, \mathrm{Re}=\frac{p_{f}cd}{\mu_{f}},\\
&M=\sqrt{\frac{\sigma}{\mu}}B_{0}d,, =\frac{\left(T-T_{0}\right)}{\left(T_{1}-T_{0}\right)}, Br=Ec.\mathrm{Pr},\\
&Q=\frac{Q_{0}d^{2}}{k_{f}(T_{1}-T_{0})}, Rn=\frac{16\sigma^{*}T_{0}^{3}}{3k^{*}k_{f}}, \mathrm{Pr}=\frac{\mu_{f}c_{f}}{k_{0}}, \mathrm{Ec}=\frac{c^{2}}{c_{f}\left(T_{1}-T_{0}\right)},\\
&E_{1}=\frac{-\tau d^{3}}{\lambda^{3}\mu_{f}c}, E_{2}=\frac{m_{1}cd^{3}}{\lambda^{3}\mu_{f}}, E_{3}=\frac{cd^{3}}{\lambda^{2}\mu_{f}}, \overline{q}_{r}=-\frac{16T_{0}^{3}}{3k^{*}}\frac{\partial\overline{T}}{\partial\overline{y}},\\
&h=\frac{\overline{h}}{d}=1+nx+\varepsilon\sin 2\pi(x-t).
\end{aligned}
\tag{7.6}
$$

The non-dimensional governing equations are demonstrated for low Reynolds numbers, and long-wavelength approximations are as follows:

$$\frac{\partial p}{\partial x}=A_{1}\left(1+\frac{1}{\beta}\right)\frac{\partial^{3}\psi}{\partial y^{3}}-\frac{M^{2}A_{4}}{\left(1+\left(A_{4}m\right)^{2}\right)}\cos^{2}\eta\frac{\partial\psi}{\partial y}-\frac{A_{1}}{Da}\frac{\partial\psi}{\partial y}, \tag{7.7}$$

$$\frac{\partial p}{\partial y}=0, \tag{7.8}$$

$$(A_{3}+Rn)\frac{\partial^{2}\theta}{\partial y^{2}}+BrA_{1}\left(1+\frac{1}{\beta}\right)\left(\frac{\partial^{2}\psi}{\partial y^{2}}\right)+Q=0, \tag{7.9}$$

where,

$$A_{1}=\frac{\mu_{hnf}}{\mu_{f}}=\frac{1}{\left(1-\left(\phi_{Cu}+\phi_{Al_{2}O_{3}}\right)\right)^{\frac{5}{2}}}, \tag{7.10}$$

$$A_{2}=\frac{(\rho\beta)_{hnf}}{(\rho\beta)_{f}}=1-\left(\phi_{Cu}+\phi_{Al_{2}O_{3}}\right)+\frac{\left(\phi_{Cu}(\rho\beta)_{p1}+\phi_{Al_{2}O_{3}}(\rho\beta)_{p2}\right)}{(\rho\beta)_{f}}, \tag{7.11}$$

$$A_3 = \frac{k_{hnf}}{k_f} = \frac{k_2(s-1)k_{nf} - (s-1)\phi_{Al_2O_3}(k_{nf} - k_2)}{k_2(s-1)k_{nf} + \phi_{Al_2O_3}(k_{nf} - k_2)} \frac{k_1(s-1)k_f - (s-1)\phi_{Cu}(k_f - k_1)}{k_1(s-1)k_f + \phi_{Cu}(k_f - k_1)}, \tag{7.12}$$

$$A_4 = \frac{\sigma_{hnf}}{\sigma_f} = 1 + \frac{3\left(\dfrac{\phi_{Cu}\sigma_{p1} + \phi_{Al_2O_3}\sigma_{p2}}{\sigma_f} - (\phi_{Cu} + \phi_{Al_2O_3})\right)}{\left(\dfrac{\phi_{Cu}\sigma_{p1} + \phi_{Al_2O_3}\sigma_{p2}}{(\phi_{Cu} + \phi_{Al_2O_3})\sigma_f} + 2\right) - \left(\dfrac{\phi_{Cu}\sigma_{p1} + \phi_{Al_2O_3}\sigma_{p2}}{\sigma_f} - (\phi_{Cu} + \phi_{Al_2O_3})\right)}. \tag{7.13}$$

The corresponding non-dimensional boundary constraints are obtained from Sucharitha et al. (2019) and Awais et al. (2020):

$$A_1\left(1 + \frac{1}{\beta}\right)\frac{\partial^3 \psi}{\partial y^3} - \left(\frac{M^2 A_4}{\left(1 + (A_4 m)^2\right)}\cos^2\eta + \frac{A_1}{Da}\right)\frac{\partial \psi}{\partial y} = E_1 \frac{\partial^3 h}{\partial x^3} + E_2 \frac{\partial^3 h}{\partial x \partial t^2} + E_3 \frac{\partial^2 h}{\partial x \partial t}, \; at \; y = h, \tag{7.14}$$

$$\psi = 0, \; at \; y = 0, \tag{7.15}$$

$$\frac{\partial \psi}{\partial y} = 0, \theta - \beta_2 \frac{\partial \theta}{\partial y} = 0, \; at \; y = -h, \tag{7.16}$$

$$\frac{\partial \psi}{\partial y} = 0, \; \theta + \beta_2 \frac{\partial \theta}{\partial y} = 1, \; at \; y = h. \tag{7.17}$$

Methodology

By partially differentiating equation 7.7 with respect to y and utilizing equation 7.8, it can be rewritten as follows.

$$\frac{\partial^4 \psi}{\partial y^4} - M_2^{\,2}\frac{\partial^4 \psi}{\partial y^2} = 0. \tag{7.18}$$

The exact solution of Equation (7.18) can be obtained by applying boundary conditions (7.14–7.17) as follows:

$$\psi = \frac{F_2}{M_2}\left[\frac{\sinh M_2 y}{M_2 \cosh M_2 h} - y\right]. \tag{7.19}$$

The axial velocity expression is expressed as follows:

$$u = \frac{F_2}{M_2^{\,2}}\left[\frac{\cosh M_2 y}{\cosh M_2 h} - 1\right]. \tag{7.20}$$

The homotopy perturbation approach (HPM) can reliably solve both linear and non-linear ordinary and partial differential equations, including nonlinear partial differential equations for temperature can be expressed as follows:

$$H(\theta,q) = (1-q)\ [L(\theta) - L(\theta_{10})] + q\left[L(\theta) + T_3 \sinh^2 M_2 y + T_2\right]. \tag{7.21}$$

The linear operator and the initial conditions are as follows:

$$L = \frac{\partial^2}{\partial y^2}, \quad \theta_{10} = \frac{y + \beta_2 + h}{2h + 2\beta_2}, \tag{7.22}$$

$$\frac{\partial^2 \theta_0}{\partial y^2} = 0. \tag{7.23}$$

Boundary conditions are written as follows:

$$\theta_0 + \beta_2 \frac{\partial \theta_0}{\partial y} = 1, \quad at \quad y = h, \tag{7.24}$$

$$\theta_0 - \beta_2 \frac{\partial \theta_0}{\partial y} = 0, \quad at \quad y = -h, \tag{7.25}$$

$$\frac{\partial^2 \theta_1}{\partial y^2} + T_3 \sinh^2 M_2 y + T_2 = 0. \tag{7.26}$$

Boundary conditions are as follows:

$$\theta_1 - \beta_2 \frac{\partial \theta_1}{\partial y} = 0, \quad at \quad y = -h, \tag{7.27}$$

$$\theta_1 + \beta_2 \frac{\partial \theta_1}{\partial y} = 0, \quad at \quad y = h. \tag{7.28}$$

The expression of θ in the power series form in terms of q is expressed as follows:

$$\theta = \theta_0 + q\,\theta_1 + \cdots \tag{7.29}$$

TABLE 7.2
Thermo physical values of nanoparticles

Physical properties	Blood	Cu	Al_2O_3
$\rho\left(Kgm^{-3}\right)$	*1063*	8933	3970
$C_p\left(Jkg^{-2}K^{-1}\right)$	*3594*	385	765
$k\left(Wm^{-1}K^{-1}\right)$	*0.492*	401	40
$\beta\times10^{-5}\left(K^{-1}\right)$	*1.8*	1.67	0.85
$\sigma\left(S\ m^{-1}\right)$	6.67×10^{-1}	5.96×10^{7}	3.5×10^{7}

Sources: Ali et al., 2023; Iftikhar et al., 2019; Waheed et al., 2023.

As q tends to 1, the analytical solution for the temperature equation can be expressed as:

$$\theta = \theta_0 + \theta_1. \tag{7.30}$$

$$\theta = L_1 y + L_2 + N_1 y^2 + N_3 \cosh M_2 h + L_3 h + L_4. \tag{7.31}$$

The list of expressions for each coefficient is in the Appendix of this chapter. Table 7.1 lists various nanoparticle shapes, their shape factors, and visual descriptions of each entry.

Thermophysical properties of the baseline blood fluid and the nanoparticle materials are provided in Table 7.2.

7.3 DISCUSSION OF THE RESULTS

This section explores physical flow outcomes such as fluid velocity, temperature, and streamline distributions to better understand the effect of regulatory factors on channel flow.

7.3.1 Velocity Profile

The effect of an applied magnetic field on the velocity profile related to Hartmann number M is portrayed in Figure 7.2a. The axial velocity appears to be a declining function of the Hartmann number because an increase in the Hartmann number (M) strengthens the Lorentz force's influence, which resists the body forces. The hybrid nanofluid's axial velocity spikes for the higher (K) values (see Figure 7.2b) due to the more porous medium in the fluid flow. Figure 7.2c illustrates that increasing the Hall current parameter (m) results in an enhanced velocity of the fluid flow, as the Hall parameter counteracts the Hartman number. Figure 7.2d demonstrates that the hybrid nanofluid velocity expands as the inclined magnetic field parameter's (η) value climbs. Figure 7.2e displays that the hybrid nanofluid velocity drops as the Casson fluid parameter's value hikes. Figures 7.2f and 7.2g illustrate the impact of wall characteristics on velocity profiles. A spike in E_1 and E_2 leads to an increase in velocity, owing to flexible walls reducing the resistance to fluid flow. Figure 7.2g

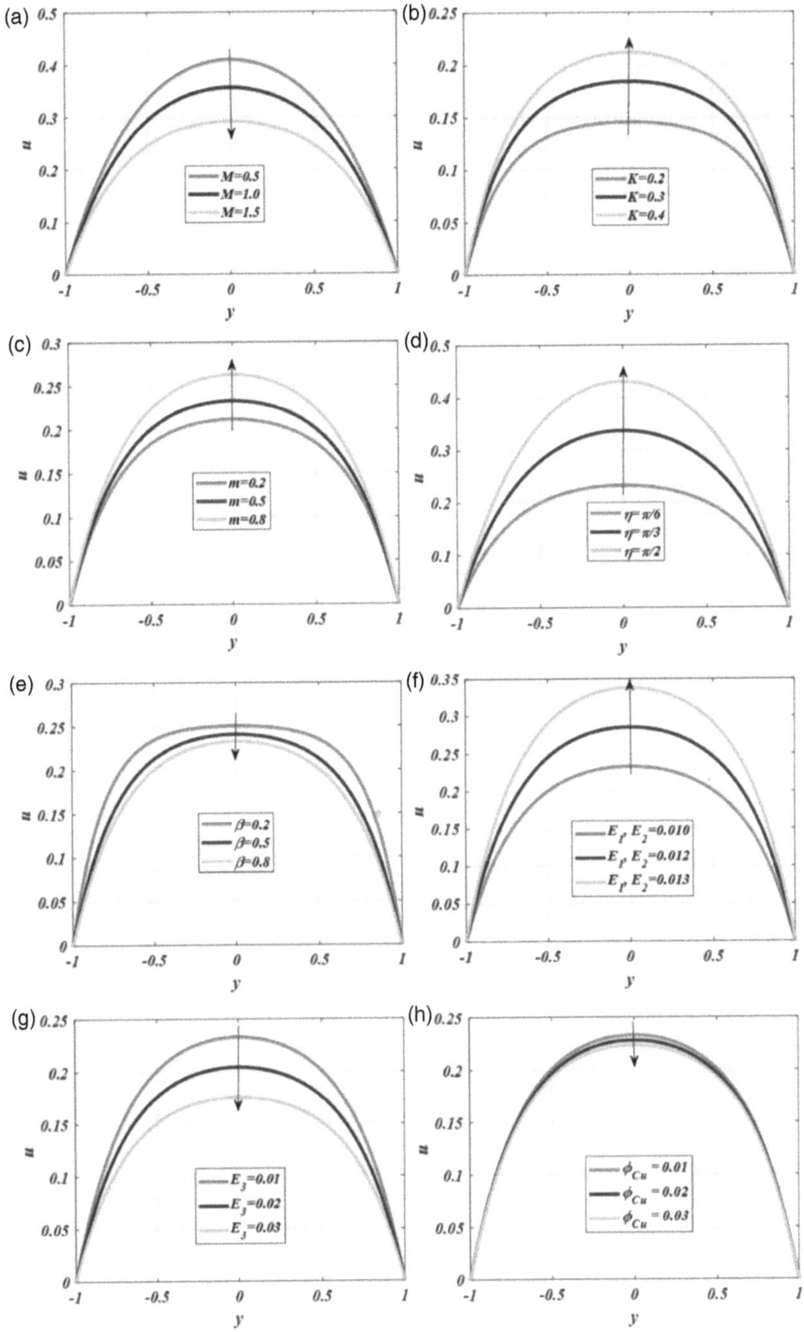

FIGURE 7.2 Effects of pertinent parameters on the velocity profile. (a) Impact of M on u. (b) Impact of K on u. (c) Impact of m on u. (d) Impact of η on u. (e) Impact of β on u. (f) Impact of E_1,E_2on u. (g).Impact of E_3 on u. (h) Impact of ϕ_{Cu} on u.

displays that as E_3 increases, the velocity profile declines; damping is a phenomenon that causes a drop in axial velocity. Moreover, an increase in the concentration of the nanoparticle (Cu) results in a greater restriction to fluid movement and lowers fluid flow velocity, as illustrated in Figure 7.2h.

7.3.2 Temperature Profile

Figure 7.3a illustrates the spike in temperature resulting from rising values of the thermal slip parameter β_2; this phenomenon occurs due to the temperature

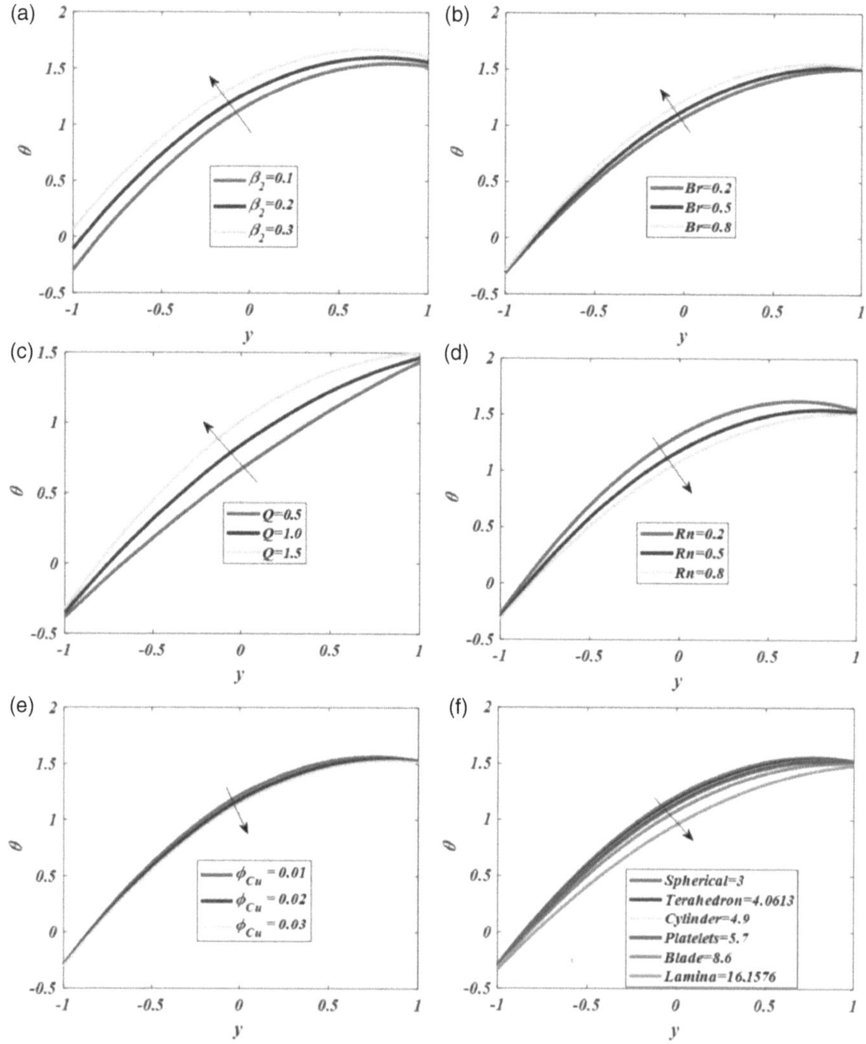

FIGURE 7.3 Effects of pertinent parameters on the temperature profile. (a) Impact of β_2 on θ. (b) Impact of Br on θ. (c) Impact of Q on θ. (d) Impact of Rn on θ. (e) Impact of ϕ_{Cu} on θ. 7.3(f) Impact of S on θ.

disparity between the fluid and the wall, leading to an increase in the fluid's temperature. The examination of Figure 7.3b reveals that the Br influences the ratio of viscous dissipation heat to molecular conduction heat; a higher value indicates less conduction heat transmission and a higher temperature elevation due to the conversion of kinetic energy into thermal energy. As seen in Figure 7.3c, the temperature profile rises as the heat source parameter (Q) value increases. The temperature profiles displayed in Figure 7.3d change as a function of the parameter Rn. As Rn increases, it causes a fall in temperature resulting from thermal radiation, which influences both the rate of heat transfer and the distribution of temperature. This reduction emerges through enhanced radiative heat loss and a thinner thermal boundary layer, leading to a drop in temperature. The concentration of the nanoparticles (ϕ_{Cu}) increases, leading to a decrease in the temperature profile, as depicted in Figure 7.3e. In Figure 7.3f, temperature profiles display that the spherical shape of the nanoparticle has more temperature relative to other shapes of nanoparticles owing to their compact shape.

7.3.3 Streamlines

During peristalsis, the bolus moves in wave-like patterns. The "splitting of streamlines" refers to the bolus's flow separating into distinct routes as it travels. This wave motion is produced by the bolus being pushed forward by peristalsis. The bolus is guided in the proper path along the digestive tract by the splitting. Figure 7.4a shows that the size of the trapping bolus increases with an increase in the Casson fluid parameter. The trapping bolus' size decreases with an increase in the permeability parameter, as illustrated in Figure 7.4b. The study examines the impact of Hartman number M on trapping, as depicted in Figure 7.4c. It demonstrates how the trapped bolus's volume rises as M grows.

7.3.4 Validation

Figure 7.5 validates the limiting scenario of the velocity distribution in the current study under the following conditions ($m = 0$, $\eta = 0$, $\Phi_{Cu} = 0$, $\Phi Al_2O_3 = 0$, and $K = \infty$). These conditions depict a simplified scenario employed to compare the findings with those published by Sucharitha et al. (2019). The graphic illustrates the alignment of the present study's findings with the results from the prior study under these specific scenarios. This validation confirms that the assumptions in the current study are reasonable and aligned with previous research.

7.4 CONCLUSIONS

This research project explores the potential industrial, engineering, and biological applications of Casson nanofluid. It examines heat radiation, viscous dissipation, and MHD peristaltic transport in porous environments. The study aims to analyze the unsteady magnetohydrodynamic flow of Casson hybrid nanofluid in porous channels, focusing on $Cu\text{-}Al_2O_3$ blood hybrid nanofluids. Some of the intriguing findings of the research are listed below:

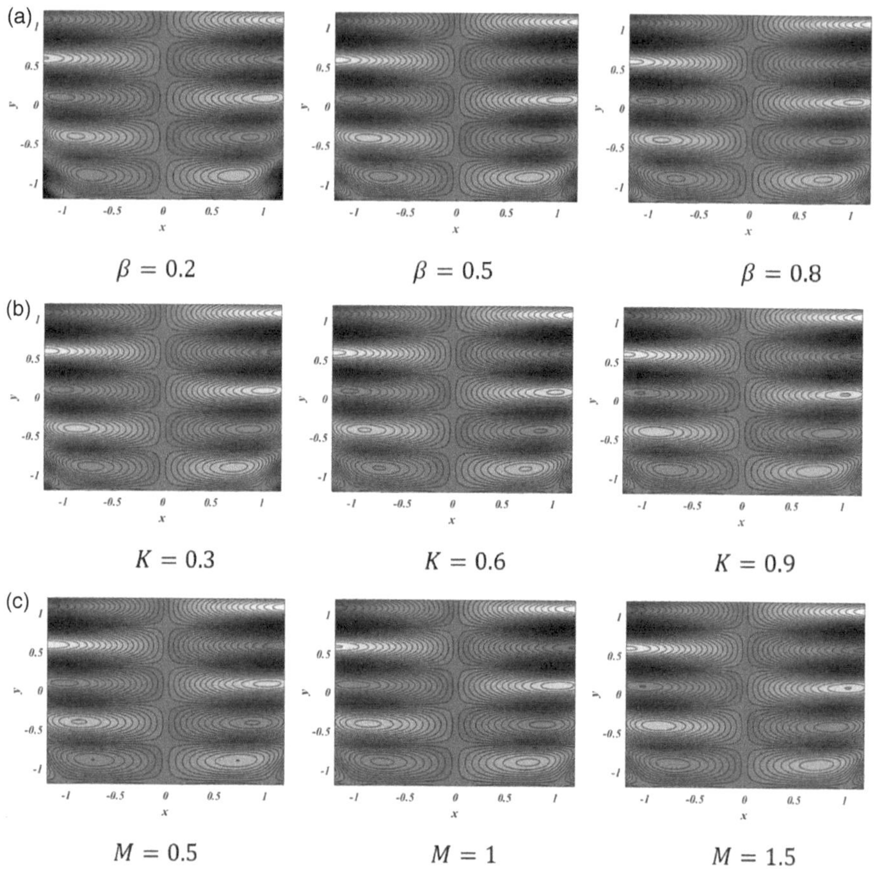

FIGURE 7.4 Effects of pertinent parameters on the streamline profile. (a) Impact of β on streamlines. (b) Impact of K on streamlines. 7.4(c) Impact of M on streamlines.

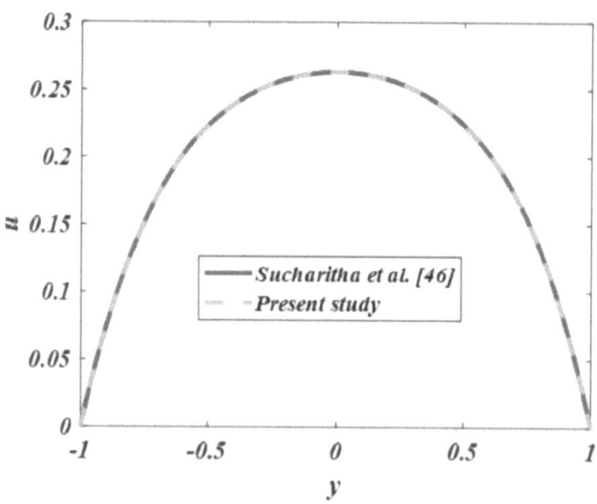

FIGURE 7.5 Validation of velocity profile.

1. A rise in the Hartmann number M seems to reduce the axial velocity because it increases the Lorentz force, which opposes body forces.
2. As the value of the Casson fluid parameter increases, the hybrid nanofluid velocity decreases.
3. A rise in E_1 and E_2 causes a surge in velocity.
4. The temperature profile decreases as the nanoparticle's volume fraction (ϕ_{Cu}) rises.
5. Temperature profiles show that spherical nanoparticles have higher temperatures compared to particles with different shapes.
6. The size of the trapping bolus reduces as the permeability parameter escalates.

APPENDIX

$$F_1 = 4\pi^2\ \varepsilon E_3 \sin 2\pi(x-t) - 8\pi^3 \varepsilon\left(E_1 + E_2\right)\cos 2\pi(x-t),$$

$$M_1 = \frac{\sqrt{A_4}}{\sqrt{1+(A_4 m)^2}} M \cos\eta + \sqrt{\frac{A_1}{Da}},$$

$$M_2 = M_1 \sqrt{\frac{1}{A_1\left(1+\frac{1}{\beta}\right)}},$$

$$F_2 = F_1 \sqrt{\frac{1}{A_1\left(1+\frac{1}{\beta}\right)}},$$

$$T_1 = \frac{Br\ A_1}{\left(A_3 + Rn\right)} \text{ Type equation here,}$$

$$T_2 = \frac{Q_0}{\left(A_3 + Rn\right)} \text{ Type equation here,}$$

$$T_3 = \frac{T_1 F_2^2}{\left(M_2^2 \cosh^2 M_2 h\right)},$$

$$L_1 = \frac{1}{\left(h + \beta_2\right)},$$

$$L_2 = \frac{1}{2},$$

$$N_1 = \left(-\frac{T_3}{4} - \frac{L_2 T_2}{2} \right),$$

$$N_2 = \left(-\frac{\beta_2 T_3}{2} - \beta_2 T_2 L_2 \right),$$

$$N_3 = \frac{T_3}{4M_1^2},$$

$$N_4 = \frac{T_3 \beta_2}{4M_2},$$

$$L_5 = N_1 h^2 + N_2 h + N_3 \cosh M_2 h + N_4 \sinh M_2 h,$$

$$L_4 = \frac{-L_5 - L_6}{2},$$

$$L_3 = \frac{L_4 + L_6}{(h + \beta_2)},$$

$$L_6 = N_1 h^2 + N_2 h + N_3 \cosh(-M_2 h) - N_4 \sinh(-M_2 h),$$

$$L_4 = \frac{-L_5 - L_6}{2},$$

$$L_3 = \frac{L_4 + L_6}{(h + \beta_2)}.$$

Nomenclature

(x, y)	coordinates in lab frame (m)
ϕ_{Cu}	the volume fraction of Cu nanoparticles
$\phi_{Al_2O_3}$	the volume fraction of Al_2O_3 nanoparticles
σ_f	the electric conductivity of the fluid $(S\,m^{-1})$

μ_f	the viscosity of the fluid $\left(kg\,s^{-1}m^{-1}\right)$
C_{pf}	specific heat of the fluid $\left(Jkg^{-2}K^{-1}\right)$
ρ_f	density of fluid $\left(Kgm^{-3}\right)$
β_f	the thermal expansion coefficient of fluid $\left(K^{-1}\right)$
K_f	the thermal conductivity of nanofluid $\left(Wm^{-1}K^{-1}\right)$
T	dimensional temperature
C_{phnf}	specific heat of hybrid nanofluid $\left(Jkg^{-2}K^{-1}\right)$
μ_{hnf}	the viscosity hybrid nanofluid $\left(kg\,s^{-1}m^{-1}\right)$
ρ_{hnf}	the density hybrid nanofluid $\left(Kgm^{-3}\right)$
β_{hnf}	the thermal expansion coefficient of hybrid nanofluid $\left(K^{-1}\right)$
K_{hnf}	the thermal conductivity of hybrid nanofluid $\left(Wm^{-1}K^{-1}\right)$
σ_{hnf}	the electric conductivity of hybrid nanofluid $\left(S\ m^{-1}\right)$
δ	wave number
Q	dimensionless heat source parameter
p	dimensionless pressure
θ	dimensionless temperature $\left(K\right)$
T_0	mean temperature
η	Inclined magnetic parameter
B_0	strength of the magnetic field
ψ	stream function
m	Hall current
Pr	Prandtl number
M	Hartman number
β_2	temperature slip parameter
Br	Brinkman number
Ec	Eckert number
Re	Reynolds number
(u,v)	velocity components in lab frame $\left(ms^{-1}\right)$
Q_0	dimensional heat generation / absorption parameter $\left(Wm^{-2}K^{-1}\right)$
p'	dimensional pressure
β	Casson fluid parameter
Rn	thermal radiation parameter
E_1, E_2, E_3	The parameters of complaint walls

REFERENCES

Abd-Alla, A.M., Abo-Dahab, S.M., Thabet, E.N., Bayones, F.S. and Abdelhafez, M.A., 2023. Heat and mass transfer in a peristaltic rotating frame Jeffrey fluid via porous medium with chemical reaction and wall properties. *Alexandria Engineering Journal*, *66*, pp.405–420. https://doi.org/10.1016/j.aej.2022.11.016

Abo-Elkhair, R.E., Bhatti, M.M. and Mekheimer, K.S., 2021. Magnetic force effects on peristaltic transport of hybrid bio-nanofluid (AuCu nanoparticles) with moderate Reynolds number: An expanding horizon. *International Communications in Heat and Mass Transfer*, *123*, p.105228. http://dx.doi.org/10.1016/j.icheatmasstransfer.2021.105228

Abraham, J.P., Sparrow, E.M. and Lovik, R.D., 2008. Unsteady, three-dimensional fluid mechanic analysis of blood flow in plaque-narrowed and plaque-freed arteries. *International Journal of Heat and Mass Transfer*, *51*(23–24), pp.5633–5641.

Afsar, H., Peiwei, G., Alshamrani, A., Alam, M.M., Hendy, A.S. and Zaky, M.A., 2023. Electroosmotically induced peristaltic flow of a hybrid nanofluid in asymmetric channel: Revolutionizing nanofluid engineering. *Case Studies in Thermal Engineering*, *52*, p.103779. http://dx.doi.org/10.1016/j.csite.2023.103779

Ajithkumar, M., Lakshminarayana, P. and Vajravelu, K., 2023. Peristaltic flow of bioconvective Ree–Eyring nanofluid through an inclined elastic channel with partial slip effects. *Journal of Applied Physics*, *134*(15), p.154701.

Akbar, N.S., Akram, J., Fiaz Hussain, M., Maraj, E.N. and Muhammad, T., 2024. Hybrid nanofluid flow and heat transfer in symmetric porous ducts with CuO nanoparticles and multi-walled carbon nanotubes under peristaltic motion. *Modern Physics Letters B*, *38*, p.2450333. https://doi.org/10.1142/S0217984924503330

Akbar, Y., Abbasi, F.M. and Shehzad, S.A., 2020. Thermal radiation and Hall effects in mixed convective peristaltic transport of nanofluid with entropy generation. *Applied Nanoscience*, *10*, pp.5421–5433. https://doi.org/10.1007/s13204-020-01446-3

Akbar, Y., Iqbal, J., Hussain, M., Khan, H. and Alotaibi, H., 2022. Peristaltic transportation of Carreau–Yasuda magneto nanofluid embedded in a porous medium with heat and mass transfer. *Waves in Random and Complex Media*, *32*(6), pp.3011–3031. https://doi.org/10.1080/17455030.2022.2036388

Ali, A., Mebarek-Oudina, F., Barman, A., Das, S. and Ismail, A.I., 2023. Peristaltic transportation of hybrid nano-blood through a ciliated micro-vessel subject to heat source and Lorentz force. *Journal of Thermal Analysis and Calorimetry*, *148*(14), pp.7059–7083. https://doi.org/10.1007/s10973-023-12217-x

Ali, N., Sajid, M., Abbas, Z. and Javed, T., 2010. Non-Newtonian fluid flow induced by peristaltic waves in a curved channel. *European Journal of Mechanics-B/Fluids*, *29*(5), pp.387–394. https://doi.org/10.1016/j.euromechflu.2010.04.002

Alqarni, A.J., Abo-Elkhair, R.E., Elsaid, E.M., Abdel-Aty, A.H. and Abdel-wahed, M.S., 2023. Effect of magnetic force and moderate Reynolds number on MHD Jeffrey hybrid nanofluid through peristaltic channel: Application of cancer treatment. *The European Physical Journal Plus*, *138*(2), pp.1–30. https://doi.org/10.1140/epjp/s13360-023-03689-9

Aslam, F. and Noreen, S., 2024. Electrokinetic peristaltic transport of Bingham-Papanastasiou fluid via porous media. *ZAMM-Journal of Applied Mathematics and Mechanics/Zeitschrift für Angewandte Mathematik und Mechanik*, *104*(2), p.e202300070. https://doi.org/10.1002/zamm.202300070

Awais, M., Kumam, P., Parveen, N., Ali, A., Shah, Z. and Thounthong, P., 2020. Slip and hall effects on peristaltic rheology of Copper-Water nanomaterial through generalized complaint walls with variable viscosity. *Frontiers in Physics*, *7*, p.249. https://doi.org/10.3389/fphy.2019.00249

El-Shehawy, E.F., El-Dabe, N.T. and El-Desoky, I.M., 2006. Slip effects on the peristaltic flow of a non-Newtonian Maxwellian fluid. *Acta mechanica*, *186*, pp.141–159. https://doi.org/10.1007/s00707-006-0343-6

Fung, Y.C. and Yih, C.S., 1968. Peristaltic transport. https://doi.org/10.1115/1.3601290

Guedri, K., Bashir, T., Abbasi, A., Farooq, W., Khan, S.U., Khan, M.I., Jameel, M. and Galal, A.M., 2022. Hall effects and entropy generation applications for peristaltic flow of modified hybrid nanofluid with electroosmosis phenomenon. *Journal of the Indian Chemical Society*, *99*(9), p.100614. https://doi.org/10.1016/j.jics.2022.100614

Hariharan, P., Seshadri, V. and Banerjee, R.K., 2008. Peristaltic transport of non-Newtonian fluid in a diverging tube with different wave forms. *Mathematical and Computer Modelling*, *48*(7–8), pp.998–1017. https://doi.org/10.1016/j.mcm.2007.10.018

Hayat, T., Amjad, S., Nisar, Z. and Alsaedi, A., 2024. Analysis for peristaltic transport of MHD hybrid nanofluid with Darcy–Forchheimer porous medium. *Modern Physics Letters B*, p.2550016. https://doi.org/10.1142/S0217984925500162

Hayat, T., Nisar, Z., Alsaedi, A. and Ahmad, B., 2021. Analysis of activation energy and entropy generation in mixed convective peristaltic transport of Sutterby nanofluid. *Journal of Thermal Analysis and Calorimetry*, *143*, pp.1867–1880. https://doi.org/10.1007/s10973-020-09969-1

Iftikhar, N., Rehman, A., Sadaf, H. and Iqbal, S., 2019. Study of Al2O3/copper–water nanoparticle shape, slip effects, and heat transfer on steady physiological delivery of MHD hybrid nanofluid. *Canadian Journal of Physics*, *97*(12), pp.1239–1252. https://doi.org/10.1139/cjp-2018-0551

Iftikhar, N. and Sadaf, H., 2023. Mathematical modelling of modified hybrid nanofluid in a peristaltic diverging tube with MHD and convective boundary conditions. *Computational Particle Mechanics*, *10*(6), pp.1477–1491. https://doi.org/10.1007/s40571-023-00553-6

Ijaz, N., Zeeshan, A., Batool, S., Bhatti, M.M. and Mekheimer, K.S., 2024. Three-Dimensional multiphase peristaltic flow through a porous medium with compliant boundary walls. *Contemporary Mathematics*, pp.4874–4889. https://doi.org/10.37256/cm.5420244994

Iqbal, J. and Abbasi, F.M., 2024. Peristaltic transport of hybrid nanofluid under the effects of thermal radiation through asymmetric curved geometry: A numerical approach. *International Journal of Modelling and Simulation*, pp.1–18. https://doi.org/10.1080/02286203.2023.2296614

Iqbal, J., Abbasi, F.M. and Shehzad, S.A., 2022. Dynamics of Hall current and entropy generation in peristaltic transport of hybrid nanofluid. *Waves in Random and Complex Media*, pp.1–26. https://doi.org/10.1080/17455030.2022.2133190

Iqbal, Z., Ahmad, I. and Khan, S.U., 2024. Peristaltic transport of magnetized Williamson hybrid nanofluid in a complex wavy channel with various shape features. *Journal of Thermal Analysis and Calorimetry*, *149*(8), pp.3699–3710. https://doi.org/10.1007/s10973-024-12966-3

Ismael, A.M., Eldabe, N.T., Abou Zeid, M.Y. and El Shabouri, S.M., 2022. Thermal micropolar and couple stresses effects on peristaltic flow of biviscosity nanofluid through a porous medium. *Scientific Reports*, *12*(1), p.16180. https://doi.org/10.1038/s41598-022-20320-6

Khazayinejad, M., Hafezi, M. and Dabir, B., 2021. Peristaltic transport of biological graphene-blood nanofluid considering inclined magnetic field and thermal radiation in a porous media. *Powder Technology*, *384*, pp.452–465. https://doi.org/10.1016/j.powtec.2021.02.036

Latham, T.W., 1966. *Fluid motions in a peristaltic pump* (Doctoral dissertation, Massachusetts Institute of Technology).

Magdy, M.M., Nasr, A.G. and Abumandour, R.M., 2024. Effect of heat transfer and slipping on MHD peristaltic flow with suspended particles. *Engineering Research Journal*, *47*(2), pp.281–305. https://doi.org/10.21608/ERJM.2024.216045.1270

Maraj, E.N., Akbar, N.S., Zehra, I., Butt, A.W. and Alghamdi, H.A., 2023. Electro-osmotically modulated magneto hydrodynamic peristaltic flow of menthol based nanofluid in a uniform channel with shape factor. *Journal of Magnetism and Magnetic Materials*, *576*, p.170774. https://doi.org/10.1016/j.jmmm.2023.170774

McCash, L.B., Akhtar, S., Nadeem, S. and Saleem, S., 2021. Entropy analysis of the peristaltic flow of hybrid nanofluid inside an elliptic duct with sinusoidally advancing boundaries. *Entropy*, *23*(6), p.732. https://doi.org/10.3390/e23060732

Mekheimer, K.S., Salem, A.M. and Zaher, A.Z., 2024. Peristatcally induced MHD slip flow in a porous medium due to a surface acoustic wavy wall. *Journal of the Egyptian Mathematical Society*, *22*(1), pp.143–151.

Mernone, A.V., Mazumdar, J.N. and Lucas, S.K., 2002. A mathematical study of peristaltic transport of a Casson fluid. *Mathematical and Computer Modelling*, *35*(7–8), pp.895–912. http://dx.doi.org/10.1016/S0895-7177(02)00058-4

Misra, J.C. and Pandey, S.K., 1999. Peristaltic transport of a non-Newtonian fluid with a peripheral layer. *International Journal of Engineering Science*, *37*(14), pp.1841–1858. https://doi.org/10.1016/S0020-7225(99)00005-1

Plourde, B.D., Vallez, L.J., Sun, B., Nelson-Cheeseman, B.B., Abraham, J.P. and Staniloae, C.S., 2016. Alterations of blood flow through arteries following atherectomy and the impact on pressure variation and velocity. *Cardiovascular Engineering and Technology*, *7*, pp.280–289.

Raghunath, K., Mohana Ramana, R., Ganteda, C., Chaurasiya, P.K., Tiwari, D., Kumar, R., Buddhi, D. and Saxena, K.K., 2024. Processing to pass unsteady MHD flow of a second-grade fluid through a porous medium in the presence of radiation absorption exhibits Diffusion thermo, hall and ion slip effects. *Advances in Materials and Processing Technologies*, *10*(2), pp.754–771. https://doi.org/10.1080/2374068X.2023.2191450

Ramamoorthy, M. and Pallavarapu, L., 2022. Second order slip flow of a conducting Jeffrey nanofluid in an inclined asymmetric porous conduit with heat and mass transfer. *Multidiscipline Modeling in Materials and Structures*, *18*(6), pp.1016–1038.

Ramki, R. and Lakshminarayana, P., 2024. Electroosmotic peristaltic transport of magneto-hydrodynamic Casson nanofluid in a non-uniform wavy porous asymmetric micro-channel. *Modern Physics Letters B*, p.2550043. https://doi.org/10.1142/S0217984925500435

Riaz, A., Zeeshan, A., Bhatti, M.M. and Ellahi, R., 2020. Peristaltic propulsion of Jeffrey nano-liquid and heat transfer through a symmetrical duct with moving walls in a porous medium. *Physica A: Statistical Mechanics and its Applications*, *545*, p.123788. https://doi.org/10.1016/j.physa.2019.123788

Siddiqui, A.M. and Schwarz, W.H., 1994. Peristaltic flow of a second-order fluid in tubes. *Journal of Non-Newtonian Fluid Mechanics*, *53*, pp.257–284. https://doi.org/10.1016/0377-0257(94)85052-6

Srivastava, L.M. and Srivastava, V.P., 1984. Peristaltic transport of blood: Casson model—II. *Journal of Biomechanics*, *17*(11), pp.821–829. https://doi.org/10.1016/0021-9290(84)90140-4

Srivastava, L.M. and Srivastava, V.P., 1985. Peristaltic transport of a non-Newtonian fluid: applications to the vas deferens and small intestine. *Annals of Biomedical Engineering*, *13*, pp.137–153. https://doi.org/10.1007/bf02584235

Sucharitha, G., Lakshminarayana, P. and Sandeep, N., 2019. MHD and cross diffusion effects on peristaltic flow of a Casson nanofluid in a duct. In *Applied Mathematics and Scientific Computing: International Conference on Advances in Mathematical Sciences, Vellore, India, December 2017-Volume II* (pp. 191–201). Springer International Publishing.

Tanveer, A., Ashraf, M.B. and Masood, M., 2024. Entropy analysis of peristaltic flow over curved channel under the impact of MHD and convective conditions. *Numerical Heat*

Transfer, Part B: Fundamentals, *85*(1), pp.45–57. https://doi.org/10.1080/10407790.2023.2224507

Thirunavukarasan, K. and Sucharitha, G., 2025. Thermal performance and MHD peristaltic flow of hybrid nanofluid (Au-Ta/Blood) in an asymmetric conduit with electro-osmosis and shape factor effects. *Case Studies in Thermal Engineering*, *66*, p.105708. https://doi.org/10.1016/j.csite.2024.105708

Tong, Z.W., Ahmed, B., Al-Khaled, K., Khan, S.U., Khan, M.I., Ahmad, S., Malik, M.Y. and Xia, W.F., 2022. Peristaltic blood transport in non-newtonian fluid confined by porous soaked tube: A numerical study through galerkin finite element technique. *Arabian Journal for Science and Engineering*, pp.1–13. https://doi.org/10.1007/s13369-021-05981-1

Vaidya, H., Mebarek-Oudina, F., Kumar, R., Rajashekhar, C., Prasad, K.V., Kalal, S. and Nisar, K.S., 2024. Magnetohydrodynamic peristaltic propulsion of Casson nanofluids with slip effects over heterogeneous rough channel. *Engineering Reports*, p.e13062. https://doi.org/10.1002/eng2.13062

Waheed, S., Noreen, S., Zahri, M. and Soufyane, A., 2023. Electrothermal transport of water conveying copper, silver and alumina nanoparticles through a vertical wavy microchannel. *Nanotechnology*, *34*(47), p.475404. https://doi.org/10.1088/1361-6528/acf2a1

8 Nanofluid Flow, Heat, and Mass Transfer at a Moving Wedge

8.1 INTRODUCTION

Viscous flow past a sharp edge is a key research area of fluid mechanics having various engineering applications as discussed, for example, in Xu (2020). During the last few decades, boundary layer flows past differently shaped geometrical bodies have been studied by multiple researchers, including Murada et al. (2022). Flow past a wedge deserves special attention as it frequently occurs in several areas like thermal insulation, heat exchangers, crude oil exploration and extraction, analysis of geothermal systems, storage of nuclear wastes, hydrodynamics, aerodynamics, magnetohydrodynamics, and many others (Kebede, 2020). Flow past a wedge has attracted the attention of researchers due to its frequent occurrence in engineering applications such as thermal insulation, heat exchangers, storage of nuclear waste, crude oil extraction, geothermal systems, and many others (Nagendramma, 2015). A lateral force is converted into a crosswise splitting force by a wedge (Hani et al., 2024). Understanding the behaviors of flow and heat transport due to a wedge offers valuable insights into its applications in various technological and engineering processes (Garalleh, 2024). In the context of coastal currents and naval engineering, wedges play a significant role and deserve special attention (Xu, 2020). Falkner Skans solutions also emerge from wedge flow – simplifying the solution to the governing equations.

Several researchers considered flow past a wedge under different physical situations, including conditions with heat transfer. Yih (1998) reported the effects of suction/blowing on flow past a wedge with a uniform heat flux. Ishak et al. (2008) examined micropolar fluid flow past a wedge with a uniform surface temperature. Khan et al. (2013) analyzed mixed convective flow past a wedge and discussed the consequences of a magnetic field and convective boundary conditions. Turkyilmazoglu (2015) obtained the exact solutions for flow past a wedge in the presence of slip. Recently, Chakraborty and Janapatla (2023) reported the combined effects of variable magnetic fields, melting heat transfer, and thermal radiation on a Casson nanofluid flow past a wedge. Considering the flow of Prandtl Eyring fuzzy hybrid nanofluid past a wedge, Nadeem et al. (2023) discussed the effects of a magnetic field on the flow and heat transfer.

DOI: 10.1201/9781003571278-8

Considering nonlinear thermal radiation and entropy generation, Sindhu et al. (2024) examined three-dimensional bioconvective nanofluid flow over a wedge. They used a two-phase model to report the effects of Brownian motion and thermophoresis. Sowmya et al. (2024) conducted a three-dimensional study of the flow and heat transfer of a magnetized hybrid nanofluid passing over a wedge. Taking into account a micropolar hybrid nanofluid, conducted a computational study of flow over a wedge surface. Considering a ternary nanofluid, Kartik et al. (2024) reported the combined impacts of thermal radiation and thermophoretic deposition for flow, heat, and mass transfer over a wedge. All of the above-mentioned researchers investigated the flow over a static wedge. But in the real world, situations occur where flow past a moving wedge plays a vital role and such type of flow has wider technological applications. Inspired by the importance of flow past a moving wedge, an attempt has been made in this chapter to reveal the heat and mass transfer characteristics of nanofluid flow across a moving wedge in the presence of a magnetic field and with activation energy which has not yet been addressed by any researcher.

8.2 FORCED CONVECTION FLOW PAST A MOVING WEDGE

Fluid flow characteristics past a moving wedge have a bearing on several engineering processes (Khan and Pop, 2017). Filtration processes, a porous stretching wedge is generally used, which necessitates the investigation of electrically conducting fluid flow, maintaining the fluid temperature, which makes it a potential insulating fluid (). Considering a power-law fluid model, Postelnicu and Pop (2011) inspected the Falkner-Skan boundary layer flow past a stretching wedge. Nagendramma and Sarojamma (2015) reported the effects of a magnetic field on flow, heat, and mass transfer past a stretching wedge in the presence of convection. Ullah et al. (2016) inspected Casson fluid flow past a moving wedge in the presence of a magnetic field. Raju et al. (2016) examined MHD Casson fluid flow over a moving wedge filled with gyrotactic microorganisms and reported the effects of Brownian motion, thermophoresis, and thermal radiation. The combined effects of Brownian motion and thermophoresis on Casson fluid flow over a wedge were reported by Ullah et al. (2018). With a Williamson fluid, Hussain et al. (2024) examined the MHD mixed convective flow over a stretching wedge in the presence of Ohmic dissipation and thermal radiation. They showed that due to higher values of the Biot number, the fluid velocity and temperature both increased.

Conventional heat transfer fluids are often replaced by nanofluids due to their superior heat transport capabilities (). The ability of a fluid to transfer heat plays an important role in manufacturing industries, as the quality of the final products and the production costs often depend on the rate of heat transport (Kebede et al., 2020).

In order to improve the thermal transport ability of fluids, nanoparticles are often added. The suspension of nanoparticles into a base fluid gives rise to "nanofluids" (Kebede et al., 2020). Motivated by the widespread uses of nanofluids and the importance of flow past a moving wedge, many researchers have investigated this type of flow problem. Using a single-phase fluid model, Yacob et al. (2011) examined nanofluid flow past a stretching wedge. They reported a comparatively higher skin friction coefficient and Nusselt number for Cu-water nanofluid than for the based fluid.

Considering Soret and Dufour effects, Kashmani et al. (2017) examined nanofluid flow past a moving wedge, and they observed an increase in the rate of heat transfer with the Soret number. In contrast, an opposite trend was noted for mass transfer. With the Keller box method, Mahdy and Chamkha (2018) obtained numerical solutions for a two-phase flow of tangent hyperbolic nanofluids across a stretched wedge in a porous medium. They showed an enhancement of horizontal velocity due to higher values of the velocity ratio parameter and the Darcy number. Using the shooting technique, Atif et al. (2019) discussed the heat and mass transfer for unsteady flow of a tangent hyperbolic nanofluid past a wedge. They found that the local Nusselt and Sherwood numbers reduced with a higher pressure gradient. Considering a ferromagnetic Casson nanofluid, Jabeen (2023) reported the effects of magnetic dipole and viscous dissipation on flow past a static/moving wedge. They noted the deceleration of velocity and temperature fields due to an increase in magnetic field strength and unsteadiness. Kulkarni and Shankar (2024) explored the behavior of the mixed convective flow of a Williamson nanofluid due to a stretching/shrinking wedge in the presence of chemical reactions. They obtained dual solutions for the skin friction coefficient for higher values of a Williamson parameter. They showed that the fluid velocity increased due to higher values of the combined convection parameter. Using a ternary hybrid nanofluid, Sharma and Jindal (2024) explored flow past a stretching wedge and presented the reverse flow analysis and the consequences of transpiration on flow and heat transfer. They highlighted the efficiency of transpiration for controlling boundary layer separation caused by the adverse pressure gradient.

The features of an electrically conducting liquid in the presence of a magnetic field are the subject matter of magnetohydrodynamics (MHD), and it has vast applications in many fields, including engineering, astrophysics, geophysics, and manufacturing. Additionally, MHD is also used for plasma confinement and in projects involving thermonuclear energy generation (see Jafar et al., 2013). Recently, fluid flow due to a stretching porous wedge in the presence of a magnetic field has found a wide range of applications, including MHD energy generators, cooling operations, oil beds, nuclear reactors, plasmas, and heat exchangers (Reddy et al., 2024).

Considering nonlinear thermal radiation, Khan et al. (2017) examined nanofluid flow due to a nonlinearly stretching/shrinking wedge in the presence of a magnetic field; they used a two phase nanofluid model to report the effects of Brownian motion and thermophoresis. Waini et al. (2020) inspected hybrid nanofluid flow past a permeable stretching/shrinking wedge in the presence of a magnetic field. They found that the wedge angle parameter was critical for determining a solution. Ali et al. (2021) explored the effects of a magnetic field on nanofluid flow past a permeable stretched wedge. Habib et al. (2022) reported the effects of a magnetic field on a Prandtl nanofluid flow past a moving wedge. They used the Keller box method to find numerical solutions to the problem.

The energy necessary to start a chemical process is termed the activation energy (see Kalaivanan et al., 2020). Chemically reactive flows have applications in water emulsion, liquid metal filtration, chemical engineering, oil reservoir, thermal magnetic flux devices, heat exchangers, heat transfer in porous media, geothermal engineering, nuclear reactors cooling, food processing, fusion control, chemical reaction species, casting, metallurgy, and many others (Mishra et al., 2023). Considering a

Casson nanofluid, Zaib et al. (2017) reported the effects of activation energy on flow past a wedge. Considering bioconvection with gyrotactic microorganisms, Mondal and Pal (2022) explored the combined impacts of a binary chemical reaction and activation energy on mixed convective nanofluid flow past a stretching wedge in a porous medium. Ali et al. (2022) examined the effects of activation energy and radiation on an MHD unsteady flow of a non-Newtonian nanofluid past a wedge. They reported that the local rate of heat transport diminished due to higher values of the thermophoresis parameter, and higher activation energy caused a decrease in the Sherwood number. Using a suspension of aluminum alloy particles in water, Rekha et al. (2022) explored the effects of activation energy on hybrid nanofluid flow past a plate, a wedge, and a cone. Considering the bioconvective flow of Williamson nanofluid over a stretched Riga wedge, Mandal et al. (2023) reported the combined influences of activation energy, magnetic field, slip, and suction. Considering suction/injection, Ohmic heating, and radiation, Hussain et al. (2024) explored magneto-cross Eyring-Powell flow over a stretched wedge in the presence of activation energy. A declining temperature profile was noted due to improved values of the mass transfer parameter.

Motivated by the importance of nanofluid flow due to a moving wedge, an attempt has been made here to reveal the flow, heat, and mass transfer characteristics of nanofluid flow due to a moving wedge in the presence of a magnetic field and activation energy.

8.3 FORMULATION OF THE PROBLEM AND SOLUTION PROCEDURE

Consider steady two-dimensional forced convective flow of a nanofluid with heat transfer over a moving wedge in a viscous, incompressible nanofluid in the presence of a magnetic field of variable strength $B(x) = B_0\ (x/L)^{(m-1)/L}$, B_0 being a constant. Arrhenius activation energy is also considered. The free stream velocity far away from the wedge is $U_e(x) = U_\infty\ (x/L)^m$ for $m \le 1$ where L is a characteristic length and m is the wedge angle parameter related to the included angle $\pi\beta$ by $m = \beta / (2-\beta)$, β is the Hartee pressure gradient. The wedge is moving with a velocity $U_w\ (x) = R\ U_e\ (x)$, where R is the velocity ratio parameter. The coordinate system is chosen so that the x-axis is along the wedge surface and the y-axis is normal to the wedge surface (see Figure 8.1).

Here, $m = 0$ and $\beta\ = 0$ corresponds to a horizontal surface, whereas $m = 1$ and $\beta = 1$, the wedge angle π , which indicates that the wedge surface becomes vertical. Similarly, $m = 1/3$ and $\beta = 1/2$ indicates that the wedge angle is $\pi / 2$.

The governing equations for such type of flow are given by

$$\frac{\partial u}{\partial x}+\frac{\partial \upsilon}{\partial y}=0, \tag{8.1}$$

$$u\frac{\partial u}{\partial x}+\upsilon\frac{\partial u}{\partial y}=u_e(x)\frac{du_e(x)}{dx}+\nu\frac{\partial^2 u}{\partial y^2}-\frac{\nu B^2(x)}{\rho}(u-u_e), \tag{8.2}$$

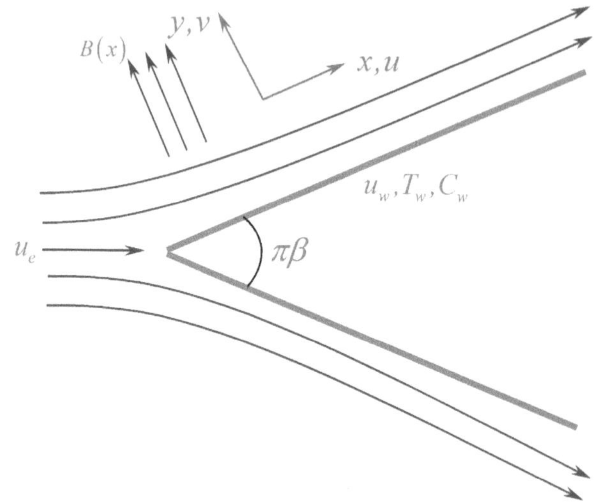

FIGURE 8.1 Geometry of the flow problem.

$$u\frac{\partial T}{\partial x}+v\frac{\partial T}{\partial y}=\alpha\frac{\partial^2 T}{\partial y^2}+\tau\left[\frac{\partial T}{\partial y}\frac{\partial C}{\partial y}+\frac{D_T}{T_\infty}\left(\frac{\partial T}{\partial y}\right)^2\right], \tag{8.3}$$

$$u\frac{\partial C}{\partial x}+v\frac{\partial C}{\partial y}=D_B\frac{\partial^2 C}{\partial y^2}+\frac{D_T}{T_\infty}\left(\frac{\partial T}{\partial y}\right)^2-K_r^2(x)\left(\frac{T}{T_\infty}\right)^n \exp\left(-\frac{E_q}{K_1 T}\right)(C-C_\infty), \tag{8.4}$$

where$v=\mu/\rho$, μ is the coefficient of viscosity of the fluid, T and C represent, respectively, the temperature and concentration of the nanofluid. The thermal diffusivity is $\alpha\kappa$, τ is the ratio of heat capacitances of the nanoparticles and the base fluid, D_B and D_T represent diffusion coefficients due to Brownian motion and thermophoresis, respectively. The symbol $K_r(x) = K_0(x/L)^{(m-1)/2}$, ($K_0$ is a constant) is the variable reaction speed, n is the fitted rate, exponent, E_q is the coefficient of activation energy, and K_I K_1 is the Boltzmann constant. The symbols T_∞ and C_∞ are, respectively, the temperature and concentration of the free stream.

The boundary conditions for this problem are:

$$u = Ru_e(x),\ v=0, T=T_w, C=C_w \text{ at } y=0, \tag{8.5}$$

$$u \to u_e(x), T\to T_\infty, C\to C_\infty \text{ as } y\to\infty. \tag{8.6}$$

Here, T_w and C_w are the temperature and concentration of the fluid at the surface of the wedge.

Let us now introduce the following relations for u and v as

$$u=\frac{\partial\psi}{\partial y},\ v=-\frac{\partial\psi}{\partial x}, \tag{8.7}$$

where ψ is the stream function.

Let us also introduce the following variables:

$$\eta=y\sqrt{\frac{(m+1)U_\infty}{2\nu L^m}}\,x^{\frac{m-1}{2}},\ \psi=\sqrt{\frac{2\nu U_\infty}{(m+1)L^m}}\,x^{\frac{m+1}{2}}f(\eta),$$

$$\theta(\eta)=\frac{T-T_\infty}{T_w-T_\infty},\ \phi=\frac{C-C_\infty}{C_w-C_\infty}. \tag{8.8}$$

Using the above relations, Equations (8.2)-(8.4) take the following forms:

$$f'''+ff''+\frac{2m}{m+1}(1-f'^2)-M\left(2-\frac{2m}{m+1}\right)(f'-1)=0, \tag{8.9}$$

$$\frac{1}{Pr}\theta''+f\theta'+Nb\,\theta'\phi'+Nt\,\theta'^2=0, \tag{8.10}$$

$$\frac{1}{Sc}\left(\phi''+\frac{Nb}{Nt}\theta''\right)+f\phi'-\Omega\left(2-\frac{2m}{m+1}\right)(\theta\gamma+1)^n\exp\left(-\frac{E}{\theta\gamma+1}\right)\phi=0. \tag{8.11}$$

To simply the expressions, multiple dimensionless numbers are defined as shown here:

$M=\sigma B_0^2L/\rho\, U_\infty$,
$Pr=\nu/\alpha$,
$Nb=\tau D_B/\nu\,(C_w-C_\infty)$,
$Nt=\tau D_T/\nu T_\infty\,(T_w-T_\infty)$,
$Sc=\nu/D_B$,
$\Omega=K_0^2\,L/U_\infty$,
$\gamma=(T_w-T_\infty)/T_\infty$.

The boundary conditions finally become:

$$f'(\eta)=R, f(\eta)=0, \theta(\eta)=1,\ \phi(\eta)=1 \text{ at } \eta=0, \tag{8.12}$$

$$f'(\eta)=1, \theta(\eta)\to 0, \phi(\eta)\to 0 \text{ as } \eta\to\infty. \tag{8.13}$$

TABLE 8.1
The values of $f''(0)$ for various values of m when $R = 0$ and $M = 0$

m	Yih (1998)	Yacob et al. (2011)	Rosenhead (1963)	Watanabe (1990)	Present study
0	0.469600	0.469600		0.469600	0.469600
1/11	0.654979	0.655000		0.654980	0.654994
0.2	0.802125	0.802100		0.802130	0.802126
1/3	0.927653	0.927700		0.927650	0.927680
0.5		1.038900			1.038903
1	1.232588	1.232600	1.232588		1.232588

The wedge moves faster than the free stream when $R > 1$ and slower than the free stream when $R < 1$. If $R < 0$, the wedge and fluid move in opposite directions, whereas $R = 0$ presents the case of a static wedge.

The above equations (8.9)–(8.11) are solved numerically using the boundary conditions (8.12)–(8.13) with the help of bvp4c code from the MATLAB® software suite.

For the verification of the numerical code, the results obtained are compared with the available results found in the open literature and found in excellent agreement for ordinary viscous fluid flow past a static wedge in the absence of a magnetic field (i.e., for $R = 0$, $M = 0$) which can be seen from Table 8.1. This excellent agreement provides confidence in the results. An additional observation from Table 8.1 is that the velocity gradient rises at the surface with an increase in the wedge angle parameter, m.

8.4 RESULTS AND DISCUSSION

Figures 8.2a–c illustrate the effects of the velocity ratio parameter R on velocity, temperature, and concentration profiles. An increase in R leads to a rise of velocity for $R > 0$ (see Figure 8.2a). As no forces oppose the flow across the surface of the wedge, the fluid velocity increases. Moreover, it is observed that the fluid velocity is higher for higher values of wedge angle parameter m, and the velocity boundary layer becomes thin when R approaches 1. With an increase in the wedge angle parameter m, the applied pressure on the fluid increases, which promotes the flow. Figure 8.2b displays temperature profiles for increasing R for $R > 0$. The temperature reduces with the rise in velocity ratio parameter R for both $R < 1$ and $R > 1$. For $R < 1$, a higher temperature is noted for lower values of m, and *for* $R > 1$, a higher temperature is observed for larger values of the wedge angle parameter m (Figure 8.2b). Though the concentration initially rises with the rise in velocity ratio parameter R (> 0) away from the surface, the concentration reduces for rising R (for $R < 1$ and $R > 1$) (Figure 8.2c).

Figures 8.3a–c display velocity, temperature, and concentration, respectively, for variations of the velocity ratio parameter R. Here, fluid velocity reduces with the decreasing values of R (see Figure 8.3a). But the fluid velocity is higher for lower

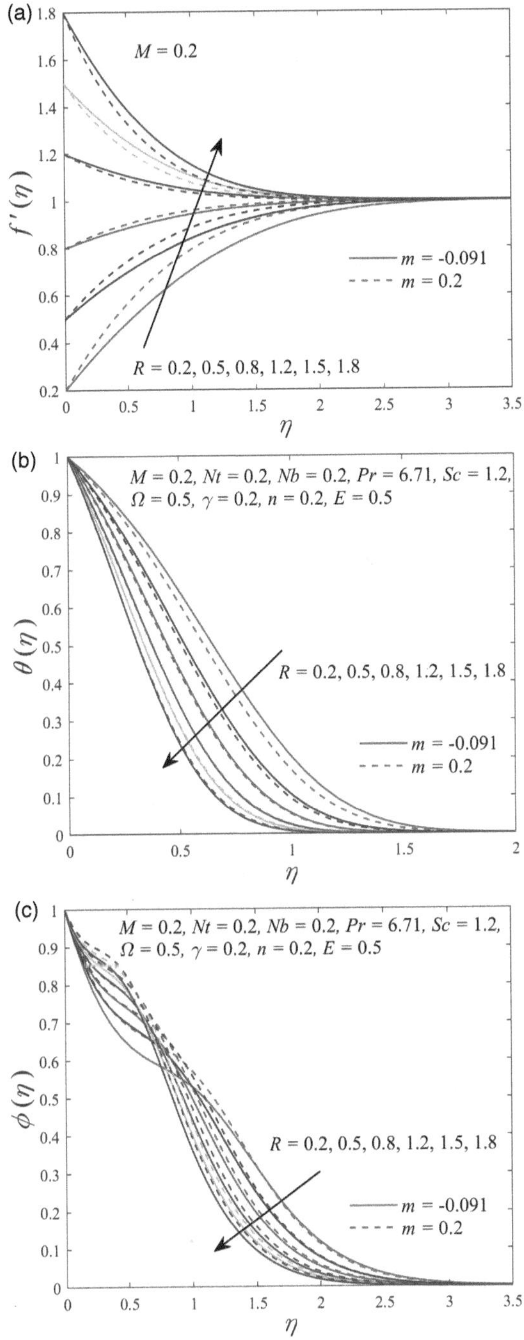

FIGURE 8.2 (a) Velocity, (b) temperature, and (c) concentration profiles for the variations of velocity ratio parameter, R, when $R > 0$.

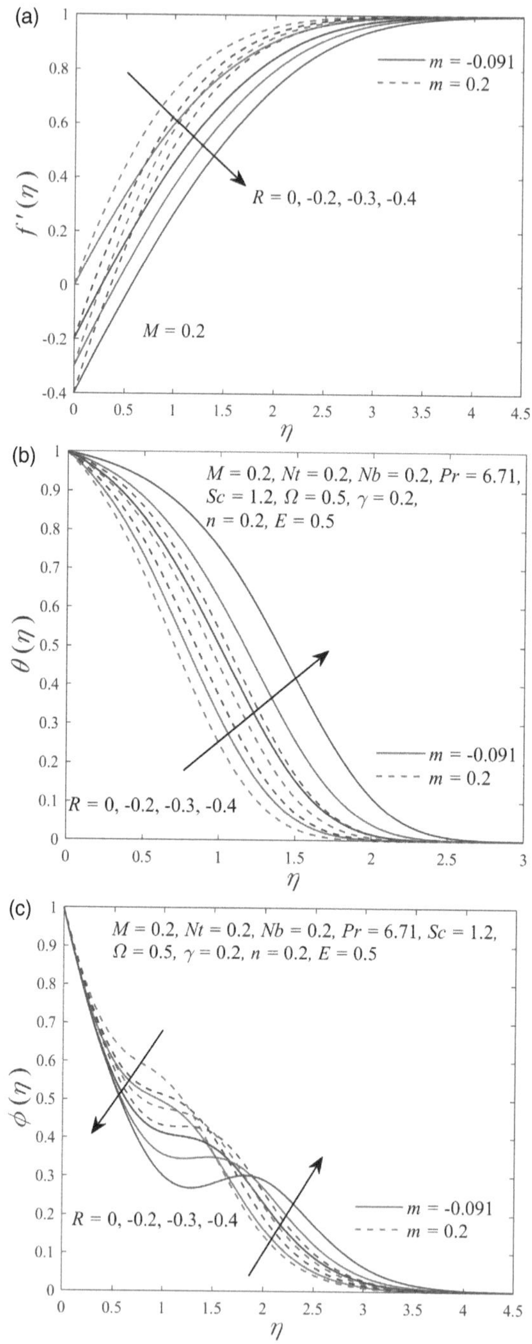

FIGURE 8.3 (a) Velocity, (b) temperature, and (c) concentration profiles for variations of velocity ratio parameter R when $R < 0$.

values of the wedge angle parameter m. $R = 0$ corresponds to a static wedge, whereas $R < 0$ occurs when the wedge and the fluid move in opposite directions. For uniform movement of the nanofluid with the wedge surface, flow velocity rises with the rise in magnitude of R. Figure 8.3(b) displays the behavior of the temperature field for $m = 0.2$ and $m = -0.091$ for a static wedge ($R = 0$) and when the wedge and fluid move in opposite directions (i.e., for $R \leq 0$). The fluid temperature rises with the increasing magnitude of R ($R < 0$) for $m = 0.2$ and also for $m = -0.091$ (Figure 8.3b). From Figure 8.3c, the concentration decreases initially with increasing magnitude of R ($R < 0$), but later, the behavior reverses.

The fluid velocity rises with the magnetic parameter M for $R > 0$ (see Figure 8.4a) and also for $R < 0$ (Figure 8.5a). With the rise in magnetic parameter M, the Lorentz force increases, and the fluid accelerates. Higher velocity is observed for larger values of the wedge angle parameter m for $R > 0$ (Figure 8.4a) and $R < 0$ (Figure 8.5a). Temperature reduces with rising values of the magnetic parameter M for $R > 0$ (Figure 8.4b) and for $R < 0$ (Figure 8.5b). Though the concentration rises initially with the rise in magnetic parameter, this behavior later reverses for growing values of M when $R > 0$ (Figure 8.4c) and when $R < 0$ (Figure 8.5c).

The temperature rises with the Brownian motion parameter Nb for $R > 0$ (see Figure 8.6a) and for $R < 0$ (Figure 8.7a). Elevation in fluid temperature and heat transfer is observed as the molecular collisions and the kinetic energy both increase due to the rise in Nb. But the concentration reduces as Nb increases when $R > 0$ (Figure 8.6b) and when $R < 0$ (Figure 8.7b).

Increasing temperature is observed with a rise in the thermophoresis parameter Nt both for $R > 0$ (Figure 8.8a) and $R < 0$ (Figure 8.9a). The concentration initially diminishes with the rise in thermophoresis parameter Nt near the surface, both for $R > 0$ (Figure 8.8b) and for $R < 0$ (Figure 8.9b). However, the increasing values of the thermophoresis parameter Nt cause a rise in concentration both for $R > 0$ (Figure 8.8b) and for $R < 0$ (Figure 8.9b). The hot fluid moves away from the surface of the wedge due to a rise in thermophoresis parameter Nt, and as a result, the temperature within the boundary layer increases.

The temperature reduces with the rise in Prandtl number Pr for $R > 0$ (Figure 8.10a). Concentrations initially rise with Pr but later diminish when $R > 0$ (Figure 8.10b).

For $R < 0$, i.e., the fluid and the wedge move in opposite directions, and the temperature initially rises with the rise in Pr but later diminishes for further Pr increases (Figure 8.11(a)). Interesting behavior in the concentration field is observed in the case of $R < 0$. In this case, though the concentration rises in the middle portion, it decreases in the initial and final regions with Pr (Figure 8.11b).

Rising concentrations are observed with a rise in activation energy parameter E for $R > 0$ (Figure 8.12a) and for $R < 0$ (Figure 8.13a), which can be understood from Eq. (8.11). But the concentration diminishes with the rise in the chemical reaction parameter Ω (Figure 8.12b) for $R > 0$ and $R < 0$, as seen in Figures 8.12b and 8.13b. Here, the chemical reaction parameter Ω improves the rate of mass transfer and reduces the concentration of the nanofluid.

The concentration decreases with a rise in Schmidt number Sc both for $R > 0$ (Figure 8.14a) and $R < 0$ (Figure 8.14b) as the molecular diffusivity diminishes.

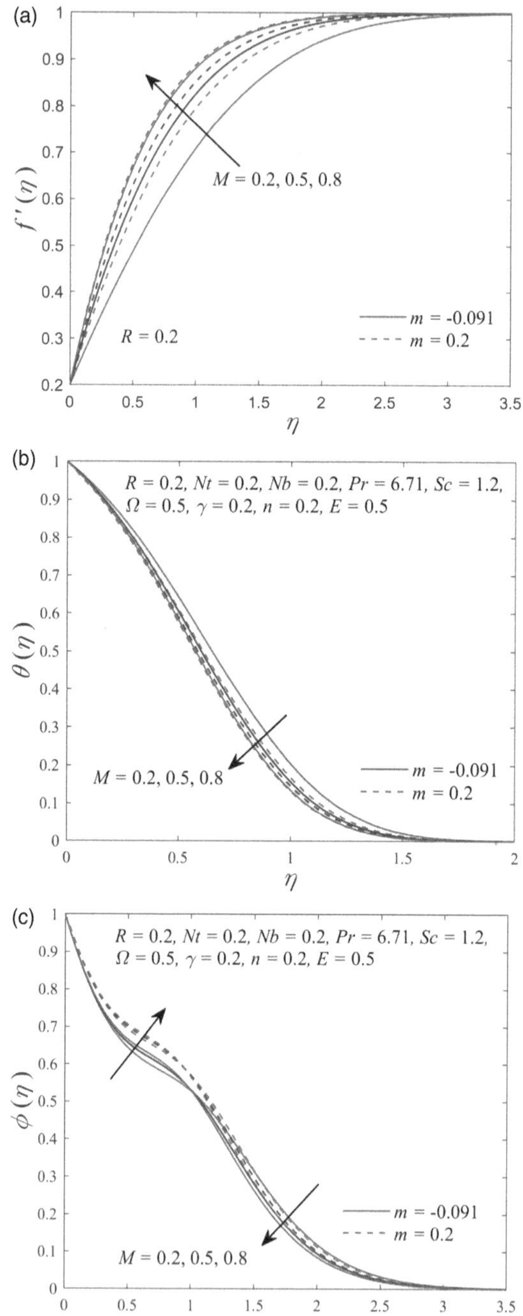

FIGURE 8.4 (a) Velocity, (b) temperature, and (c) concentration profiles for the variations of magnetic parameter M when $R > 0$.

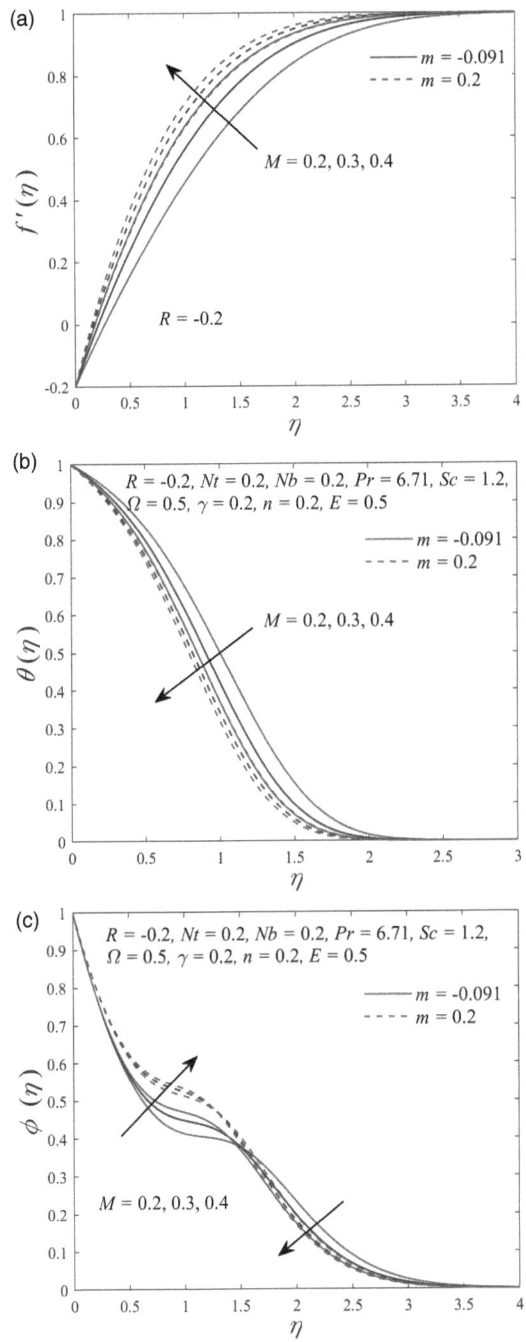

FIGURE 8.5 (a) Velocity, (b) temperature, and (c) concentration profiles for the variations of magnetic parameter M when $R < 0$.

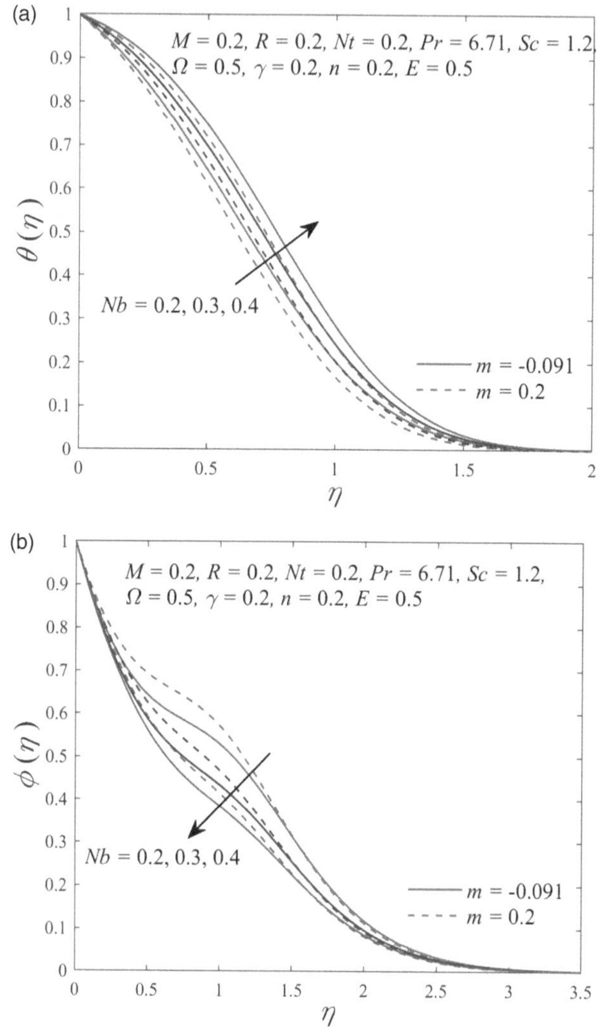

FIGURE 8.6 (a) Temperature and (b) concentration profiles for the variations of the Brownian motion parameter, Nb, when $R > 0$.

8.5 CONCLUSIONS

In this chapter, the flow behavior and heat mass transport features for MHD nanofluid flow due to a moving wedge have been analyzed. Three cases have been considered: (i) when the fluid and the wedge move in the same direction ($R > 0$), (ii) when the fluid and the wedge move in opposite directions ($R < 0$), and (iii) when the wedge is not moving ($R = 0$), only the fluid moves. A two-phase model for nanofluids has been considered for exploring the effects of Brownian motion and thermophoresis. In addition, the impact of activation energy has been reported. The results of this study

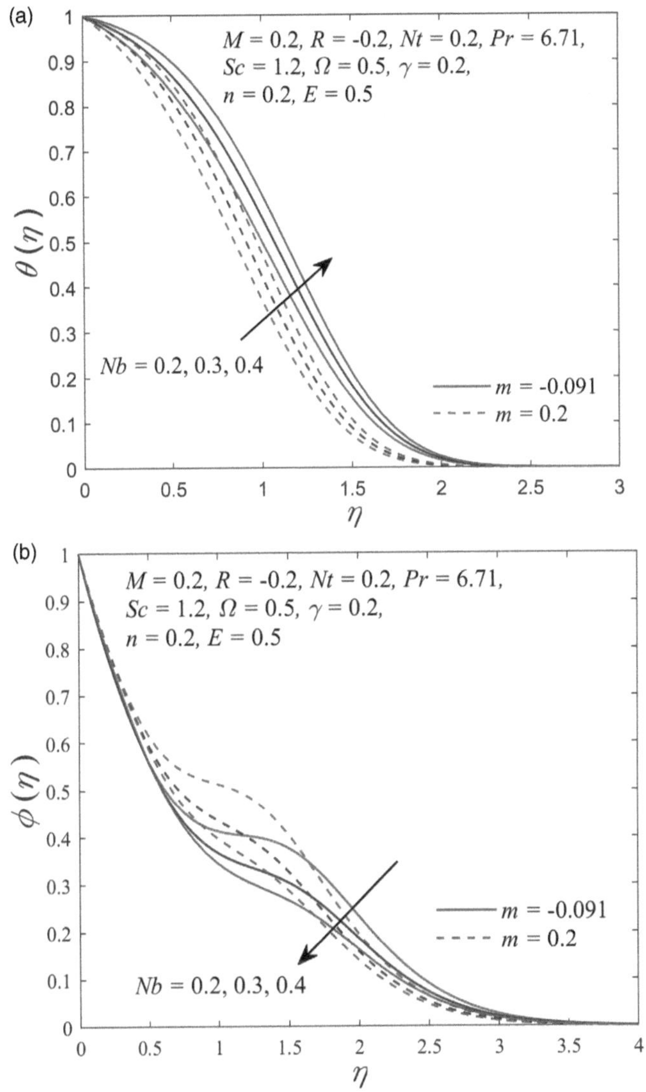

FIGURE 8.7 (a) Temperature and (b) concentration profiles for the variations of Brownian motion parameter *Nb* when $R < 0$.

display fascinating behaviors of flow, heat, and mass transfer, which deserve more detailed analysis in the future for some other physical situations.

Nomenclature

D_B	Brownian diffusion coefficient
D_T	thermophoresis diffusion coefficient
E	activation energy parameter

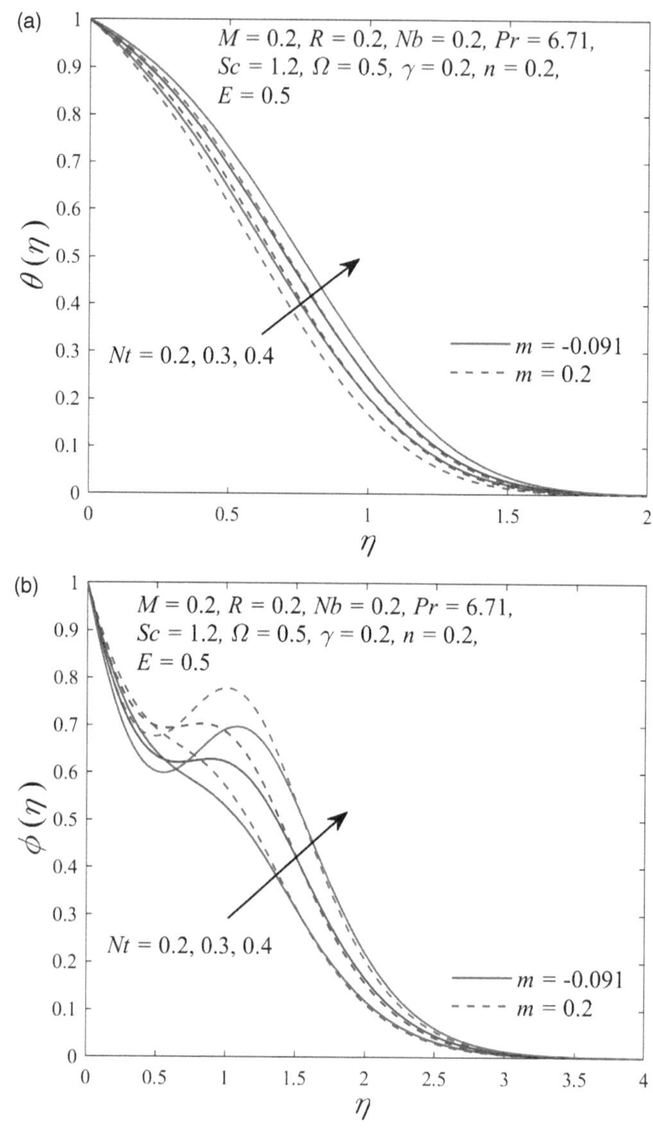

FIGURE 8.8 (a) Temperature and (b) concentration profiles for the variations of Brownian motion parameter *Nt* when $R > 0$.

K_l	Boltzmann constant
Nb	Brownian motion parameter
M	magnetic parameter
Nt	thermophoresis parameter
Pr	Prandtl number
Sc	Schimdt number

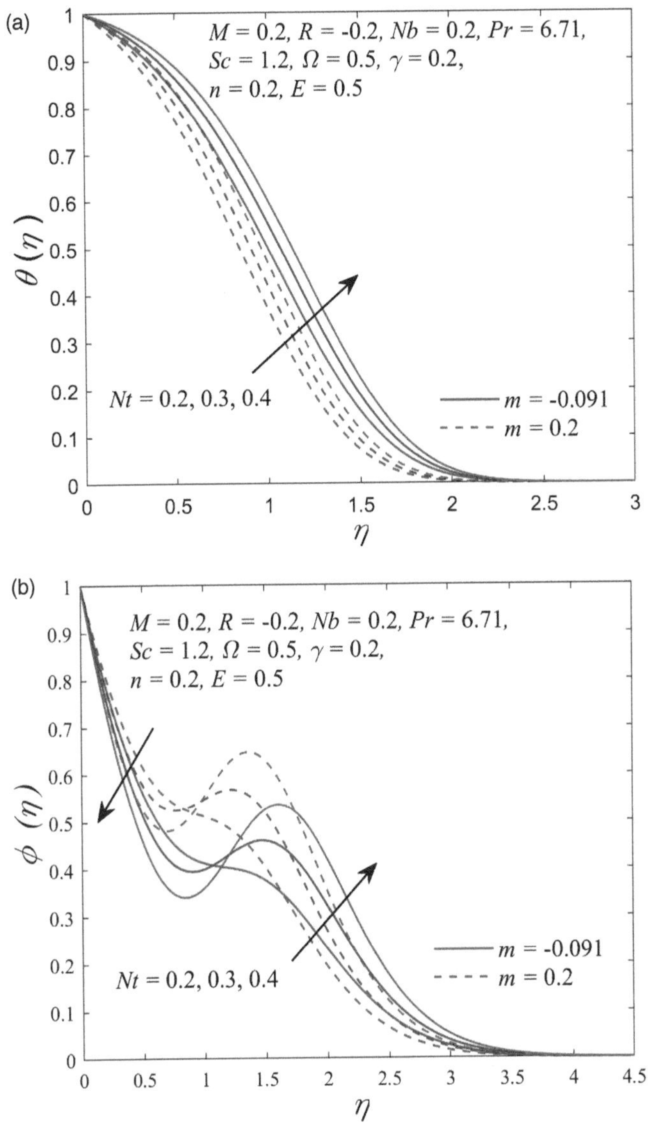

FIGURE 8.9 (a) Temperature and (b) concentration profiles for the variations of Brownian motion parameter Nt when $R < 0$.

T	temperature of the liquid (K)
T_∞	free stream temperature (K)
T_w	surface temperature (K)
u, v	velocity component in x, y direction (m/s)
α	thermal diffusivity
η	similarity variable

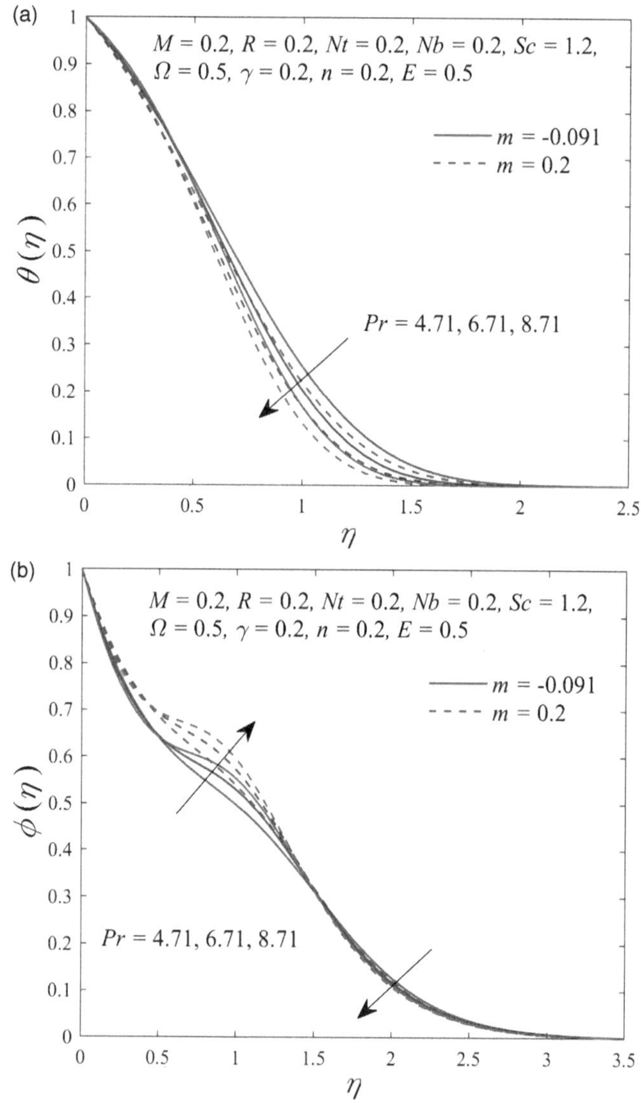

FIGURE 8.10 (a) Temperature and (b) concentration profiles for variable values of Pr when $R > 0$.

ν	kinematic viscosity (m^2/s)
θ	non-dimensional fluid temperature
ρ	liquid density (kg m^{-3})
Ω	Chemical reaction parameter
γ	temperature ratio parameter
ψ	stream function

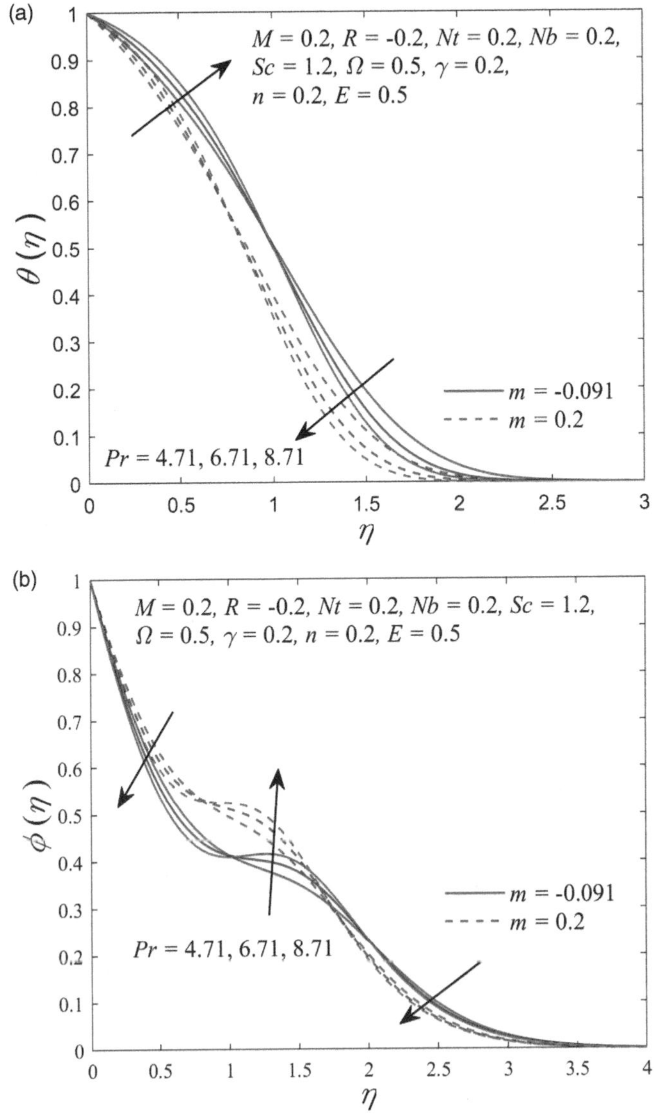

FIGURE 8.11 (a) Temperature and (b) concentration profiles for variable values of *Pr* when $R < 0$.

REFERENCES

AL Garalleh, H., 2024. Numerical simulation of the heat transport mechanism in chemically influenced ternary hybrid nanofluid flow over a wedge geometry. *Discover Applied Sciences*, *6*(9), p.449. https://doi.org/10.1007/s42452-024-06141-4

Ali, L., Ali, B., Liu, X., Iqbal, T., Zulqarnain, R.M. and Javid, M., 2022. A comparative study of unsteady MHD Falkner–Skan wedge flow for non-Newtonian nanofluids considering thermal radiation and activation energy. *Chinese Journal of Physics*, *77*, pp.1625–1638.

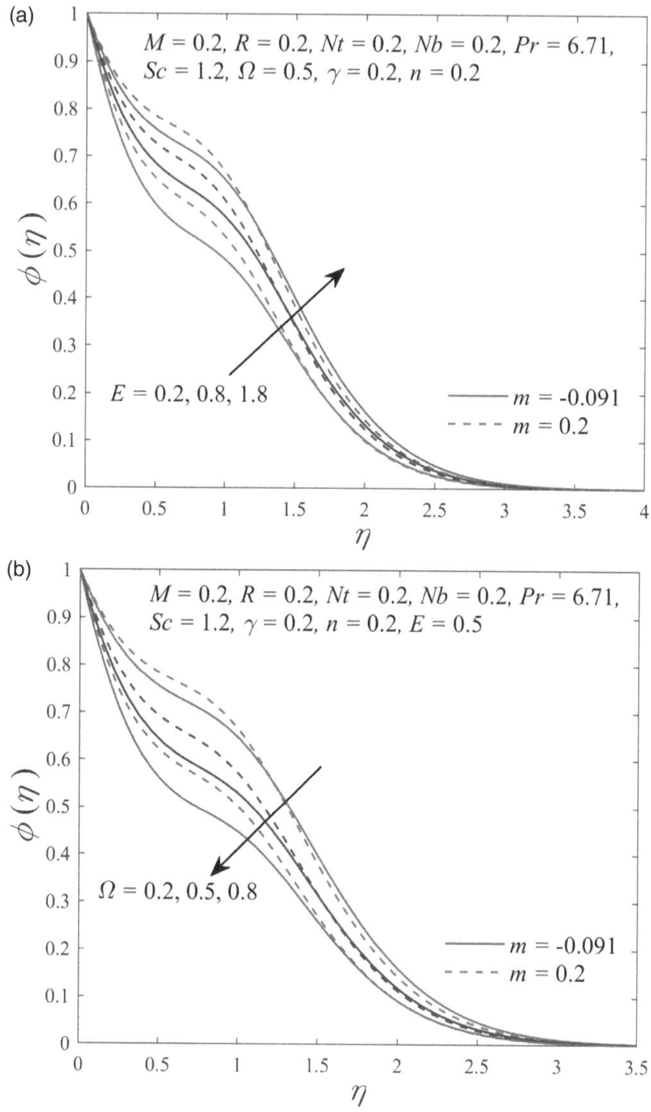

FIGURE 8.12 Concentration profiles for variable values of (a) E and (b) Ω when $R > 0$.

Ali, M., Nasrin, R. and Alim, M.A., 2021. Analysis of boundary layer nanofluid flow over a stretching permeable wedge-shaped surface with magnetic effect. *Journal of Naval Architecture and Marine Engineering*, *18*(1), pp.11–24.

Atif, S.M., Hussain, S. and Sagheer, M., 2019. Heat and mass transfer analysis of time-dependent tangent hyperbolic nanofluid flow past a wedge. *Physics Letters A*, *383*(11), pp.1187–1198.

Chakraborty, A. and Janapatla, P., 2023. Study of Casson nanofluid flow over a wedge with variable magnetic effect, thermal radiation, and melting effects. *Numerical Heat Transfer, Part B: Fundamentals*, *85*, pp.1437–1459..

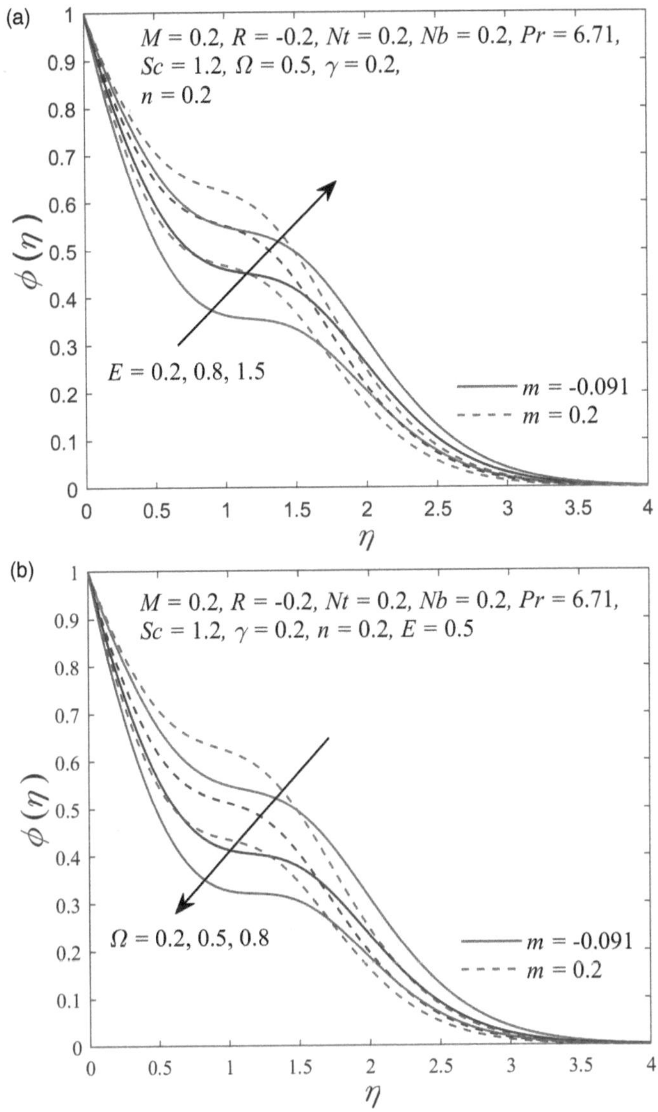

FIGURE 8.13 Concentration profiles for variable values of (a) E and (b) Ω when $R < 0$.

Habib, D., Salamat, N., Abdal, S.H.S. and Ali, B., 2022. Numerical investigation for MHD Prandtl nanofluid transportation due to a moving wedge: Keller box approach. *International Communications in Heat and Mass Transfer*, *135*, p.106141.

Hani, U., Ali, M. and Alam, M.S., 2024. MHD boundary layer micropolar fluid flow over a stretching wedge surface: Thermophoresis and Brownian motion effect. *Journal of Thermal Engineering*, *10*(2), pp.330–349.

Hussain, M., Ali, A., Ranjha, Q.A., Ahmad, I. and Anwar, M.S., 2024. Radiative magneto-cross Eyring-Powell flow with activation energy past porous stretching wedge

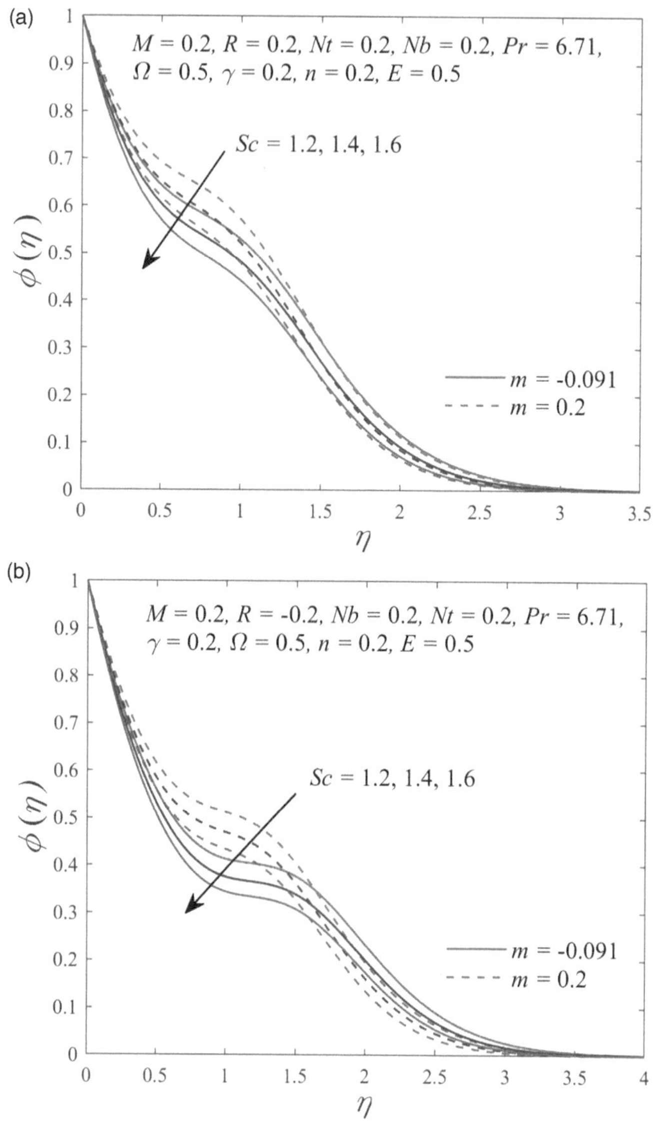

FIGURE 8.14 Concentration profiles for variable values of *Sc* for (a) $R > 0$ and (b) $R < 0$.

considering suction/injection and ohmic heating effect. *Numerical Heat Transfer, Part B: Fundamentals*, *85*(7), pp.867–882.

Hussain, M., Lubna, A., Ashraf, M., Anwar, M.S., Ranjha, Q.A. and Ali, A., 2024. Ohmically dissipated MHD mixed convective flow of Williamson fluid over a penetrable stretching convective wedge with thermal radiations. *Numerical Heat Transfer, Part B: Fundamentals*, *85*(8), pp.1026–1040.

Ishak, A., Nazar, R. and Pop, I., 2008. MHD boundary-layer flow of a micropolar fluid past a wedge with variable wall temperature. *Acta Mechanica*, *196*, pp.75–86.

Jabeen, K., 2023. Impact of magnetic dipole on unsteady ferromagnetic Casson nanofluid flow over a static/moving wedge in the presence of viscous dissipation. *Numerical Heat Transfer, Part B: Fundamentals*, *84*(6), pp.773–793.

Jafar, K., Nazar, R., Ishak, A. and Pop, I., 2013. MHD boundary layer flow due to a moving wedge in a parallel stream with the induced magnetic field. *Boundary Value Problems*, *2013*, pp.1–14.

Kalaivanan, R., Ganesh, N.V. and Al-Mdallal, Q.M., 2020. An investigation on Arrhenius activation energy of second grade nanofluid flow with active and passive control of nanomaterials. *Case Studies in Thermal Engineering*, *22*, p.100774.

Karthik, K., Madhukesh, J.K., Kiran, S., Nagaraja, K.V., Prasannakumara, B.C. and Fehmi, G., (in press). Impacts of thermophoretic deposition and thermal radiation on heat and mass transfer analysis of ternary nanofluid flow across a wedge. *International Journal of Modelling and Simulation*. doi: 10.1080/02286203.2023.2298234

Kebede, T., Haile, E., Awgichew, G. and Walelign, T., 2020. Heat and mass transfer analysis in unsteady flow of tangent hyperbolic nanofluid over a moving wedge with buoyancy and dissipation effects. *Heliyon*, *6*(4). https://doi.org/10.1016/j.heliyon.2020.e03776

Khan, U., Ahmed, N., Mohyud-Din, S.T. and Bin-Mohsin, B., 2017. Nonlinear radiation effects on MHD flow of nanofluid over a nonlinearly stretching/shrinking wedge. *Neural Computing and Applications*, *28*, pp.2041–2050.

Khan, W. A. and Pop, I., 2013. The boundary layer flows past a moving wedge in a nanofluid. *Mathematical Problems in Engineering,* 2013, p.637285. https://doi.org/10.1155/2013/637285

Kulkarni, M. and Shankar, H.F., (in press). Numerical investigation of mixed convective Williamson nanofluid flow over a stretching/shrinking wedge in the presence of chemical reaction parameter and liquid hydrogen diffusion. *Numerical Heat Transfer, Part A: Applications*. doi: 10.1080/10407782.2024.2335550.

Mahdy, A. and Chamkha, A.J., 2018. Unsteady MHD boundary layer flow of tangent hyperbolic two-phase nanofluid of moving stretched porous wedge. *International Journal of Numerical Methods for Heat & Fluid Flow*, *28*(11), pp.2567–2580.

Mandal, R.K., Maiti, H. and Nandy, S.K., 2023. Bioconvective MHD flow of Williamson nanofluid past an expandable Riga wedge in the presence of activation energy, mass suction and velocity slip. *Numerical Heat Transfer, Part A: Applications*, 86, pp.368–397.

Md Kasmani, R., Sivasankaran, S., Bhuvaneswari, M. and Hussein, A.K., 2017. Analytical and numerical study on convection of nanofluid past a moving wedge with Soret and Dufour effects. *International Journal of Numerical Methods for Heat & Fluid Flow*, *27*(10), pp.2333–2354.

Mishra, N.K., Anwar, S., Kumam, P., Seangwattana, T., Bilal, M. and Saeed, A., 2023. Numerical investigation of chemically reacting jet flow of hybrid nanofluid under the significances of bio-active mixers and chemical reaction. *Heliyon*, *9*(7), p.e17678.

Mondal, KS. and Pal, D., 2022. Gyrotactic mixed bioconvection flow of a nanofluid over a stretching wedge embedded in a porous media in the presence of binary chemical reaction and activation energy. *International Journal of Ambient Energy*, *43*(1), pp.3443–3453.

Murad, N.M., Rawi, N.A., Shafie, S., Lim, Y.J. and Mahat, R., 2022. Unsteady Falkner-Skan flow of hybrid nanofluid over a nonlinear moving wedge. *Malaysian Journal of Fundamental and Applied Sciences*, *18*(1), pp.116–123.

Nadeem, M., Siddique, I., Bilal, M. and Anjum, K., 2023. Numerical study of MHD Prandtl Eyring fuzzy hybrid nanofluid flow over a wedge. *Numerical Heat Transfer, Part A: Applications*, *85*, pp.4328–4344..

Nagendramma, V., Sreelakshmi, K. and Sarojamma, G., 2015. MHD heat and mass transfer flow over a stretching wedge with convective boundary conditions and thermophoresis. *Procedia Engineering*, *127*, pp.963–969.

Postelnicu, A. and Pop, I., 2011. Falkner–Skan boundary layer flow of a power-law fluid past a stretching wedge. *Applied Mathematics and Computation*, *217*(9), pp.4359–4368.

Raju, C.S.K., Hoque, M.M. and Sivasankar, T., 2016. The radiative flow of Casson fluid over a moving wedge filled with gyrotactic microorganisms. *Advanced Powder Technology*, *28*(2), pp.575–583.

Reddy, M.V., Ajithkumar, M., Lone, S.A., Ali, F., Lakshminarayana, P. and Saeed, A., 2024. Magneto-Williamson nanofluid flow past a wedge with activation energy: Buongiorno model. *Advances in Mechanical Engineering*, *16*(1), p.16878132231223027.

Rekha, M.B., Sarris, I.E., Madhukesh, J.K., Raghunatha, K.R. and Prasannakumara, B.C., 2022. Activation energy impact on flow of AA7072-AA7075/Water-Based hybrid nanofluid through a cone, wedge and plate. *Micromachines*, *13*(2), p.302.

Rosenhead, L. and Talbot, L., 1963. *Laminar Boundary Layer*. Oxford University Press.

Sharma, K. and Jindal, R., 2024. Transpiration-induced convective heat transfer in ternary hybrid nanofluid flow due to dual stretching on wedge: Reverse flow analysis. *Numerical Heat Transfer, Part A: Applications*, pp.1–13. https://doi.org/10.1080/10407782.2024.2383400

Sindhu, S., Vajravelu, K. and Sowmya, G., (in press). Three-dimensional bioconvection in a nanofluid film over a wedge surface with entropy generation and nonlinear radiation: A revised Buongiorno model. *International Journal of Modelling and Simulation*, doi: 10/1080.02286203.2024.2420649.

Sowmya, G., Vajravelu, K. and Sindhu, S., (in press). Magnetized hybrid nanofluid flow across a wedge: A computational three-dimensional study. *Numerical Heat Transfer, Part A: Applications*. doi: 10.1080/10407782.2024.2343038

Turkyilmazoglu, M., 2015. Slip flow and heat transfer over a specific wedge: an exactly solvable Falkner–Skan equation. *Journal of Engineering Mathematics*, *92*(1), pp.73–81.

Ullah, I., Khan, I. and Shafie, S., 2016. Hydromagnetic Falkner-Skan flow of Casson fluid past a moving wedge with heat transfer. *Alexandria Engineering Journal*, *55*(3), pp.2139–2148.

Ullah, I., Shafie, S., Khan, I. and Hsiao, K.L., 2018. Brownian diffusion and thermophoresis mechanisms in Casson fluid over a moving wedge. *Results in Physics*, *9*, pp.183–194.

Umavathi, J.C. and Shamshuddin, M.D., (in press). Computational micropolar model of hybrid nanofluid flow across a wedge. *Numerical Heat Transfer, Part A: Applications*, doi: 10.1080/10407782.2024.2368273

Waini, I., Ishak, A. and Pop, I., 2020. MHD flow and heat transfer of a hybrid nanofluid past a permeable stretching/shrinking wedge. *Applied Mathematics and Mechanics*, *41*(3), pp.507–520.

Watanabe, T., 1990. Thermal boundary layers over a wedge with uniform suction or injection in forced flow. *Acta Mechanica*, *83*(3), pp.119–126.

Xu, L., 2020. Accelerating viscous flow past a wedge. *Physics of Fluids*, *32*(1). https://doi.org/10.1063/1.5125456

Yacob, N.A., Ishak, A. and Pop, I., 2011. Falkner–Skan problem for a static or moving wedge in nanofluids. *International Journal of Thermal Sciences*, *50*(2), pp.133–139.

Yih, K.A., 1998. Uniform suction/blowing effect on forced convection about a wedge: uniform heat flux. *Acta Mechanica*, *128*(3), pp.173–181.

Zaib, A., Rashidi, M.M., Chamkha, A.J. and Bhattacharyya, K., 2017. Numerical solution of second law analysis for MHD Casson nanofluid past a wedge with activation energy and binary chemical reaction. *International Journal of Numerical Methods for Heat & Fluid Flow*, *27*(12), pp.2816–2834.

9 Concluding Remarks and Future Areas of Research

In Chapter 1, we established a foundational understanding of nanofluid and hybrid nanofluid flows, their properties, development, and applications. We began with Choi and Eastman's seminal work in 1995 that later spawned this large and vibrant research theme in the scientific community. This chapter discusses nanofluid preparation methods including both one-step and two-step synthesis, stabilization, characterization, and the importance of nanoparticle dispersion within the base fluid. Specific case studies and recent research publications are used to illustrate the real-world application of this technology.

Next, in Chapter 2, we include the effects of radiation, chemical reactions, and MHD processes on nanofluid flow over a porous moving sheet. We explore how these phenomena influence heat and mass transfer at the surface when combined with heat sources and sinks or with viscous dissipation. Among the nanoparticles included are Titanium dioxide (TiO_2) in water – a combination that is relevant to various real-world applications.

This chapter develops a mathematical framework where the governing differential equations are articulated for the conservation of mass, momentum, heat, and species. Similarity transformations of the boundary layers convert the dimensional differential equations to dimensionless linear equations that can be solved by numerical methods using the MATLAB® bvp5c solver. This solution approach is validated against other researchers' findings and is shown to calculate velocity, temperature, and concentration profiles accurately.

Chapter 3 provides an analysis of heat and mass transfer for nanofluid flow over a stretching surface. With widespread industrial applications in areas such as polymer processing, metalwork, and thin film technology, we encounter various scenarios, including linear, nonlinear, exponential, and unsteady stretching. The chapter incorporates buoyancy, variations in the Prandtl number, Brownian motion, thermophoresis, and Lewis numbers and quantifies their impact on velocity, temperature, and concentration distributions. It was found that heat transfer depends on these dimensionless parameters, and they can be set to optimize cooling performance at a stretching surface.

With nonlinear motion, we show that the velocity decreases while temperature and concentrations generally increase with the non-linear stretching parameter. We

DOI: 10.1201/9781003571278-9

also show that passive control of nanoparticle flux has implications in flow control technologies. For exponential stretching, there is a significant dependence of flow and thermal behavior on the velocity ratio, Brownian motion, thermophoresis, Prandtl and Lewis numbers. We incorporate real-world situations with nonlinear motion.

The last significant case of this chapter is the unsteady motion of a surface – a type of motion that experiences rapid deformation, such as with changes in the operating conditions. Here, the importance of thermophysical properties, particularly for a Cu-water combination, is evident in unsteady situations with thermal radiation.

In retrospect, Chapter 3 provides a comprehensive numerical and theoretical understanding of the nanofluid behavior as it flows over the surface. This information enables optimization of industrial processes.

Next, Chapter 4 investigates the mixed convective flow of hybrid nanofluids with TiO_2 and SiO_2 nanoparticles in water over a vertical stretching plate. We incorporate MHD processes, porous effects, thermal and solutal stratification, viscous dissipation, Joule heating, buoyant-driven convection, and chemical reactions. The governing equations are simplified through similarity transformations that ultimately provide linear differential equations that can be solved numerically. We show that increasing the magnetic parameter and porosity yields a decrease in the fluid velocity on account of the Lorentz forces. On the other hand, increasing the buoyancy parameter increases the velocity because of intensified convection. The hybrid nanofluid thermal behavior is sensitive to magnetic fields and viscous dissipation, which raises the temperature, whereas increased stratification reduces it. These findings provide insights into proper thermal management strategies in solar energy systems, nuclear power production, heat exchangers, and cooling techniques.

In Chapter 5, the complex phenomena of nanofluid flow caused by stretching disks are explored. We provide multiple scenarios: flow over a radially stretching disk, flow caused by a rotating and stretching disk, and unsteady flow near a stagnation location. The impact of blowing or suction, Brownian motion, and thermophoresis are included. We demonstrate that the fluid velocity, temperature, and nanoparticle concentrations are sensitive to these parameters. Increasing the stretching enhances the velocity initially but this behavior reverses itself elsewhere. Brownian motion increases temperature and decreases nanoparticle concentrations.

We also consider nanofluid flow over a rotating and radially stretching disk – a situation that introduces rotational effects, suction/blowing at the disk, and convection. We show that increases in the rotational parameter influence the velocity, temperature, and heat flux. We conclude by investigating unsteady hybrid nanofluid flow over a stretching disk near a stagnation point. Dual solutions emerge, providing insights about stability. We show that unsteadiness profoundly affects the velocity and temperature and that the stretching parameter and nanoparticle concentration strongly influence skin friction and heat transfer.

Chapter 6 explores nanofluid flow around a stretching cylinder, a physical situation with multiple real-world applications. We compare the performance of copper and silver in water with uniform surface temperatures. Here, we detail the governing differential equations of this type of boundary layer problem (mass, momentum, heat, and concentration). We incorporate the effect of the cylinder curvature through the use of a dimensionless curvature parameter. As with earlier chapters, similarity solutions are developed following the transformation of the initial nonlinear coupled differential equations to linear equations.

We show that the presence of nanoparticles significantly affects the flow and temperature profiles – increasing the nanoparticle concentration typically reduces the velocity and the temperature profiles. These changes are a result of the increase in the fluid viscosity and thermal conductivity.

Chapter 7 turns its attention to peristaltic flows with heat transfer and MHD and with a Casson nanofluid that flows through a porous channel with flexible and stretching walls. We explored Cu-Al_2O_3 nanoparticles suspended in blood, a combination that is important for multiple biomedical therapies, including targeted drug delivery, hyperthermia treatments, and diagnostic equipment enhancement. Important physical phenomena are included in this analysis, including viscous dissipation, thermal radiation, Hall currents, and inclined magnetic fields. After dimensionless transformations, the governing differential equations are simplified with exact solutions for velocity and temperature made available through the homotopy perturbation method (HPM).

We find that the presence of a magnetic field significantly reduces the axial velocity because of the Lorentz force. On the other hand, increased permeability of the medium and increased Hall currents increase the fluid velocity. We show a profound effect of wall flexibility and Casson fluid properties on the velocity. This illustrates how structure interactions can facilitate or hinder fluid motions.

Taken together, Chapter 7 provides a theoretical framework for understanding the relationship between magnetic, thermal, and structure forcings.

The last significant chapter (Chapter 8) deals with nanofluid flow over a wedge-shaped body under both static and unsteady scenarios and with magnetic fields and activation energy. We highlight the practical significance of these types of wedge flows in real-world situations such as thermal insulation, nuclear waste storage, geothermal systems, and aerodynamic applications. The result of this analysis is the provision of guidelines for optimizing heat and mass transfer processes in various applications.

Nanotechnology in thermal-fluid sciences is a major advancement in this technical field – this advancement has exploded within the scientific community and has been marked by extensive research and applications. We are hopeful that this book can serve as a resource to researchers and engineers as they develop an understanding and intuition of this subject and as they apply it in the real world.

This comprehensive exploration of nanofluid and hybrid nanofluid flows across various geometries and physical conditions highlights numerous promising directions for future research:

1. **Nanoparticle stability and dispersion**: it is important to ensure long-term stability and stable, uniform concentration of nanoparticles within the base fluid. Future studies focused on improving both of these issues by mitigating aggregation and sedimentation and, thereby, long-term performance.
2. **Hybrid nanofluid optimization**: While nanofluids are seen to offer superior thermal performance, further improvements can be made by optimizing nanoparticle combinations, concentrations, and particle shapes.

3. **Experimental validation exercises**: While the theory of nanofluid flow has matured, there is a paucity of high-quality experiments that can serve as validations to numerical solutions and can provide an avenue for improved numerical algorithms.
4. **Advanced computational methodologies**: New numerical approaches, including machine learning, artificial intelligence, and high-performance computers, can provide deeper insights into the Multiphysics interactions within these multicomponent fluids.
5. **Real-world applications**: more effort is needed to translate a solid theoretical understanding to real-world applications. Whether used in heat exchangers, nuclear reactors, biomedical applications, manufacturing, lubrication, or others, it is time to move vigorously more from theory to practice.
6. **Deeper unsteadying of magnetic fields and activation energy**: More studies on the role of these two phenomena are needed, as well as on their combined effects in complex flow situations.
7. **Thermal radiation and chemical reaction**: similar to the preceding issue, more research is needed on flow, temperature, and concentrations in the presence of radiation and reactions. This issue is particularly germane for stability analysis.
8. **Peristaltic flows**: peristalsis is a common biological flow with tremendous real-world implications. The use of hybrid nanofluids in peristaltic flows holds promise for many biomedical applications, and future research is needed to explore this promise.
9. **Unsteady and turbulent flows**: most research is predominantly focused on steady, laminar flows. Clearly unsteadiness and turbulence are inevitable in many applications and thus the deserving of study. Unsteady flow and turbulence are commonly encountered in high-performance cooling systems and rapid manufacturing processes with dynamic operating conditions.

Index

Note: Figures are indicated by *italics*.

Kolibri 360 GmbH
Ferdinand-Jühlke-Straße 7
99095 Erfurt, Deutschland
produktsicherheit@kolibri360.de